NATIVE ORCHIDS
of
Peninsular Malaysia

NATIVE ORCHIDS
of
Peninsular Malaysia

Chris K.H. Teo

TIMES BOOKS INTERNATIONAL
Singapore • Kuala Lumpur

© 1985 Times Editions Pte Ltd
Reprinted 1995

Published by Times Books International
an imprint of Times Editions Pte Ltd
Times Centre
1 New Industrial Road
Singapore 1953

Times Subang
Lot 46, Subang Hi-Tech Industrial Park
Batu Tiga
40000 Shah Alam
Selangor Darul Ehsan
Malaysia

All rights reserved. No part of this publication may be reproduced, stored in a retrieval system or transmitted, in any form or by any means, electronic, mechanical, photocopying, recording or otherwise, without the prior permission of the copyright owner.

Printed by KHL Printing Co. Pte Ltd

ISBN 981-204-568-6

CONTENTS

Preface
Introduction

THE GENERA

Acriopsis   7
Aerides   9
Anoectochilus   11
Appendicula   14
Ascocentrum   18
Arundina   20
Bulbophyllum   22
Chiloschista   32
Cleisostoma   34
Cleisomeria   40
Corymborkis   42
Cymbidium   44

Flickingeria   50
Grammatophyllum   52
Geodorum   55
Habenaria   58
Kingidium   62
Malleola   64
Paphiopedilum   66
Phalaenopsis   70
Podochilus   75
Pomatocalpa   77
Porphyrodesme   81
Renanthera   83

Renantherella   85
Rhynchostylis   87
Robiquetia   89
Smitinandia   90
Spathoglottis   92
Thecostele   97
Thrixspermum   99
Trichoglottis   101
Vandopsis   106
Vanilla   107
Ventricularia   110

Cultivation of Species Orchids   *112*

Index   *117*

PREFACE

Developing countries of the world need to progress to keep up with "advanced" nations. Often, this progress means the felling of forest or digging up of land for the harnessing of natural resources. In the process of exploiting this God-given wealth, it is inevitable that the natural flora and fauna found in these areas are also destroyed. It is a pity, that in this race to progress, few ever appreciate the need to preserve, save or conserve our natural heritage.

Orchid species form only one part of the flora that suffer this fate of extinction. At the rate this destruction is going on, it will not take long before many species become extinct. And extinction is forever!

Like it or not, this is a reality which we must face: that for future generations, many orchid species may just be something that they only read about in books.

The first step towards conservation is to know what we still have. It is only after having taken stock of what there is that we can hope to formulate a strategy to protect and conserve. One of the aims of this book is partly to satisfy this need to catalogue.

The species included in this volume are native to Peninsular Malaysia. While some of these are already well known, many others are not. Some have horticultural value, while others may have such a potential which is yet to be realised. It is also true that there are a few which are only of botanical interest. In any case, by portraying them all in colour, I hope to drive home the point that orchid species are beautiful and unique in their own ways, and are worth conserving!

There is already a good book on Malayan orchids by Pro fessor R.E. Holttum published in 1953 and revised in 1964. Since then, there have been many name changes, extensions of distributions and new species found. The learned "Father of the Malaysian Orchids" was aware of these changes and he wrote to me thus: "Various changes of names for species should be made in a revision of my book, but the publishers will not re-set it. I hope it may be possible sometime to get a supplement prepared with reference to all changes." I hope *Native Orchids of Peninsular Malaysia* will make a useful supplement to this effect.

In the attempt, I have endeavoured to be as up-to-date as I know how. Also, I have avoided using scientific jargon in the descriptions to make the text easier reading to both layman and scientist. The materials on which research was based were all fresh specimens. For this reason, I was able to see minor details which most other researchers could not observe in dried herbarium specimens. This, I believe, is one advantage which this book has over others.

One problem I faced when preparing this work was the identification of the species. I have gone through all the available references and have written to orchid authorities abroad for help. Therefore, if there are still some mistakes on my part, I stand corrected and will appreciate their being brought to my notice.

The photographs illustrating the text were all taken by me. These represent years of effort in trekking jungle paths, studying species under microscopes and experimenting with photographic technique. However, I still owe much to many orchid-enthusiast friends for having helped me in one way or another. Without their valuable support, the production of this book would not have been possible. The assistance and contribution by all my friends at Penang Botanical Gardens (including its former Director, Mr. Cheang Kok Choy) is specially acknowledged. I also extend special thanks to Mr. Gunnar Seidenfaden for providing reference literature and taking valuable time to answer my queries.

Finally, I wish to thank the Vice-Chancellor, Universiti Sains Malaysia. Y.B. Datuk Musa bin Mohamad for his kind permission to publish this book and also for providing a special research grant to study and conserve orchid species.

CTKH
School of Biological Sciences
Universiti Sains Malaysia
Penang

INTRODUCTION

THE ORCHID SPECIES

Orchids belong to the largest group of flowering plants in the world. It is estimated that about 25,000 species representing 660 genera have so far been recorded. Of these, well over 800 species covering 120 genera are said to be indigenous to Peninsular Malaysia. This number is by no means exhaustive for perhaps, there are many more new species yet to be discovered. The figure may also be inaccurate as there may be some species which are not in fact indigenous to Malaysia but have, somehow, found themselves included in the count. Information about these may have been derived from doubtful specimens. It is the author's opinion that many more changes may be expected as more research is carried out.

Grouping
Species orchids can be divided broadly into two main groups: the terrestrials and the epiphytes.

Terrestrial orchids are those that grow with their roots buried in soil or leaf debris. These roots lack the adaptive velamen layer which soaks up water and stores it for a while. They cannot withstand exposure to the air.

From the viewpoint of cultivation, terrestrial orchids in Malaysia can be broadly divided into 2 main groups.

(a) Terrestrials with limited root system. This group may have thickened stems and produce tubers. Their leaves are generally thin. This group includes the Habenaria and Goodyera tribes, the saprophytes and their allies.

(b) Terrestrials with large root systems. These plants may produce pseudobulbs or leathery leaves. This group is represented by the Paphiopedilum, Phaius, Arundina and Liparis tribes.

The latter group is rather popular with orchid growers but the former group is hardly known. Nevertheless, both are part of a fascinating group of species which is worth cultivating.

The Habenaria tribe consists of three genera: Habenaria, Peristylis and Pecteilis. These are common in the northern regions of Peninsular Malaysia. Peristylis is abundant on Penang Hill, Gunung Jerai and parts of the lowlands. Pecteilis is less common, possibly because it has been over collected because of its large and colourful flowers. One well known, and sought after species is *Pecteilis susannae*. This very impressive species is about 120 cm high and bears up to 10 flowers, with the lip about 5cm across. Unfortunately this species has become rare and difficult to find in its wild state.

Members of the Habenaria tribe generally produce tubers which represent the dormant stage of their life cycle. This stage allows the plants to survive adverse conditions. In Malaysia, the dormant stage is during the dry season while in Europe or other temperate lands it is the cold winter.

The Goodyera tribe includes many attractive species. However, they are grown not for their flowers but for their leaves. Referred to as jewel orchids, some species are common but records on them are scanty. Jewel orchids are easy to cultivate if given conditions similar to their natural habitat.

The true saprophyte appears to be poorly represented in the Malaysian flora. Only about 14 species belonging to 9 genera (e.g. Galeola, Aphyllorchis) have so far been recorded. A member of the Goodyera tribe which is a saprophyte is *Cystorchis aphylla*. In other countries the genera Dipodium and Cymbidium also have saprophytic members. The author is inclined to believe that saprophytic orchids are not as rare in Malaysia as they seem. The reason why there are so few recorded is that they are not easily detected or found. The plant lies buried underground most of the year. The only time it can be observed is when the plant pushes its inconspicuous inflorescence out of the ground. Even then, the flowers do not last too long; after a few days most of them fade. Without the flower there is little else to indicate that such an orchid exists.

A saprophyte lacks chlorophyll and is pale in colour. It does not carry out photosynthesis, i.e. does not manufacture its own food. Its leaves have, therefore, no function and are generally absent or reduced to appear as scales along the stem. The root of a saprophytic orchid is buried in the humus layer and it lives in close association with a fungus. This fungus breaks down the complex cellulose materials into simple food for the orchid. Such saprophytes are usually found in the damp and shady places of both lowland and highland forests.

The allies of the saprophyte which are found in Malaysia are Corybas, Cryptostylis and Vanilla.

The second group of terrestrial orchids with well developed root systems includes the Paphiopedilum, Phaius and Arundina. Their large showy flowers and easy cultivation make them popular among orchid growers.

Terrestrials like Arundina, Bromheadia, Spathoglottis are naturally adapted to more exposed and drier environment. As such, they are commonly found along the edges of the forest, exposed ridges or jungle clearings. The other terrestrials such as Paphiopedilum, Calanthe and Tainia require a more damp and shady habitat.

Epiphytic orchids, like the terrestrials can also be divided into two main groups: the sympodials and the monopodials. A sympodial growth or habit is one in which the stem (or pseudbulb) has a limited growth and therefore is generally short. New growth develops from the base of a pseudobulb. In this way, the plant actually creeps along the stratum on which it grows. Examples of sympodial orchids are the Bulbophyllum, Coelogyne and Dendrobium.

The monopodial epiphytic orchids are usually climbers, but there are exceptions such as the Phalaenopsis. A monopodial plant has a stem where growth occurs at its tip and grows without terminating. Trichoglottis and Ventricularia have short slender stems that climb and attach themselves to tree branches. *Papilionanthe hookerana* grows among swamp vegetation anchoring itself in wet marsh.

Habitat

Species orchids are found in all kinds of habitats, ranging from sea-level to mountain-tops. Many perch themselves on the branches and trunks of trees. Others find themselves a suitable home on rocks, either exposed to the cool breeze of the sea and mountain air or sheltered moist places along streams. All these descriptions fit the habitats of the Malaysian native orchids. The photographs in the following pages should provide some idea of natural habitats.

The distribution of an orchid species is generally limited to a particular environment. For example, *Dendrobium crumenatum* is commonly found throughout the region extending from the Philippines to India and China. On the other hand, *Calanthe vestita*, found in the area extending from Burma to New Guinea, is rather rare.

The reason that gives rise to this difference in distribution and abundance is that each species demands a certain ecological niche. It will not grow outside its set of environmental conditions. *Dendrobium crumenatum* takes to a broader ecological niche. It is not surprising, therefore, to see this native of the East, growing luxuriantly in the warm greenhouses of the western world. *Calanthe vestita*, on the other hand, survives only in areas with a pronounced dry season.

In Malaysia, areas such as Gunung Jerai, Cameron Highlands, Mount Ophir and Taman Negara are well known to botanists and orchid hunters. Each of these habitats has its own distinctive features that support a wealth of diverse flora and fauna, orchids included. The understanding of the habitat from which the species is derived is therefore important if one is to succeed in cultivating it.

Most people associate orchids with the rain forest where lush trees support an abundance of species on their trunk and branches. The truth, however, is that few people, walking through the rain forest, actually see orchids growing on trees.

Like all plants, orchids need light to grow. So, it is not likely

A clump of *Acriopsis javanica* on the trunk of a rambutan tree in a home garden.

Ventricularia tenuicaulis hanging from a branch of a tree in its natural habitat.

Dendrobium crumenatum growing on a frangipani tree in the middle of Penang City.

Podochilus growing on moist, moss-covered rocks in a hilly area.

that they can survive on the branches under the dense canopy of thick virgin forests. Where there is a break in the forest, in clearings, near streams or rivers and in ravines or gullies, orchids can be seen more readily. Indeed, this is one aspect which can be used to great advantage by the orchid hunter. When searching, it is always difficult to find the first plant of a species. But once a plant is found many more specimens come into view.

For a beginner, hunting orchids can be a hard job. He can simply walk past many orchids without seeing them, even if his eyes remain wide open. This is because he simply does not know where to look for orchids! The experienced collector would, however, study the environment and in his mind, make an educated guess as to where orchids may be found in that particular habitat.

Diversity
The variations one finds in orchid habitats equal the diversity that exists in the form, size and structure of both the plant and the flower.

The giant orchid of Malaysia, *Grammetophyllum speciosum* grows into a sturdy, bushy plant like the sugar-cane. On the other extreme there is the inconspicuous Corybas which is but a tiny plant about a few centimetres in height. In between these extremes, are various intermediates. *Phalaenopsis gigantea* has thick, succulent leaves that measure up to 60 cm in length and about 20 cm in width. The Taeniophyllum is leafless and the stem rarely 2 cm long. What are conspicuous are the green root strands. The Vanilla is a vine that can grow up to 300 m high.

Leaf structure also varies greatly. Ludisia and Anaectochilus have beautiful ornate leaves. The entire plant is herbaceous in nature. The Corymborkis has large, pleated leaves. On the onset, the plant appears like a palm and could be mistaken for one. The Bulbophyllum and Vandaceous species among others have short, rather narrow, leathery leaves. The inflorescences also show great variation. Some are only one-flowered (e.g. some species of Trichoglottis) while others have numerous flowers borne on a cluster or placed on well-spread sprays (e.g. *Rhynchostylis retusa, Porphyrodesme elongata*).

Grammetophyllum speciosum bears inflorescences which are 2–3 metres long and carry hundreds of flowers on each. The inflorescence of *Dimorphorchis* (formerly, *Arachnis*) *lowii* is unique. It is wiry, hangs down and could be up to 3 metres long. The first two or three flowers of the inflorescence are yellow while the rest are red.

The colour of the orchid flower can take any blend and shade of the spectrum of a rainbow. Flowers range in colour from yellow to pure white, bright red, mauve or orange.

The flower on the plant can be short-lived and wither soon after opening. For example, the flower of *Bromheadia finlaysoniana* opens at dawn and withers by noon. Some others like *Phalaenopsis violacea* have flowers that last for weeks.

Some species bloom very regularly and often one finds them

in flower throughout the year, as is the case of *Phalaenopsis cornu-cervi*. Some only bloom once year, while others may flower once in many years.

Orchid flowers may possess a fragrance of some kind. Examples of two extremes are found in *Vanda dearii* and the many species of Bulbophyllum. The sweet aroma that pervades the air when *Vanda dearii* comes into bloom reminds one of the world's best perfume. On the other hand, *Bulbophyllum maximum*, for example, produces a foul odour that only blue-bottle flies can appreciate.

Unity

In spite of the great diversity one finds in orchids, there is, however, one basic characteristic which all orchids possess. This is the flower structure.

An orchid flower has three sepals. Usually one is placed at the top of the flower while two others are at the sides. All of these are usually alike, but in some genera, the top sepal can be different.

There are three petals in an orchid flower. They alternate with the sepals. The bottom petal is modified and is different from the two other petals. It is called the lip or labellum. The lip is usually large and prominent; and sometimes has a rather complex shape. It is made up of three lobes: two side-lobes and one mid-lobe.

In the centre of the flower, facing the lip is the column. This column can be short or long. It carries the anther at its free-end while on the underside is a hollow cavity. This cavity is filled with a sticky substance and is the stigma. Beneath the column is the ovary and this develops into a seed pod if fertilization occurs.

The above is a generalized structure of an orchid flower. As much as one encounters great diversity in the plant-form, there is also modification and diversity in the floral structure, although the basic characteristics described above remain constant. For example, in Paphiopedilum, the top sepal is modified and is prominently large. The two side sepals are not visible from the front, but reverse the flower and one finds these two structures united to form the synsepalum. In Pecteilis the lip is provided with a long spur which contains nectar. In Bulbophyllum the lip is hinged onto the column-foot and this allows it to swing like a perfect pendulum.

In the following pages various species and genera of orchids native to Peninsular Malaysia are described and illustrated. All these go to demonstrate the rich diversity and unique creations that Mother Nature has in store for us all.

ACRIOPSIS

This genus consists of floristically interesting species. They are found growing on trees in areas from Tenasserim down through Peninsular Malaysia up till New Guinea. Holttum listed five species native to Peninsular Malaysia: *A. javanica, A. ridleyi, A. densiflora, A. indica* and *A. carrii*.

Acriopsis javanica Reinw.

Pseudobulb: is small, ovoid, and grows into dense clusters attached to host support.
Leaf: is long, narrow and thin. There are about three leaves growing from the terminal end of each pseudobulb.
Inflorescence: is long and arches over. It can be simple or branched; and carries many small flowers.
Flower: the dorsal sepal is lightly narrower than the lateral petal and it curves forward. It is pale yellow with a brown spot at the tip. The lateral petal is yellowish with a large brown spot at the tip and a brown median band. The apex of the petal is rounded. The lateral sepals are joined to form a simple structure behind the lip. It is concave as is the dorsal sepal and similarly coloured. The lip is jointed with the basal half of the column to form a narrow tube. The lip is placed about perpendicular to the column. The side-lobe is broad and spread horizontally. The mid-lobe is tongue-shaped and is not lobed. At the constriction between the mid-lobe and the side-lobe are two thin, vertical keels. They are placed parallel and close together. The entire lip has a white edge with crimson-purple in the middle and again white in the middle.
The column is straight and at a point about three-quarters of its length high arise two forward side-arms. The tip of each arm is yellow. The column terminates with a rounded edge forming a large hood covering the anther.
Acriopsis javanica is commonly found in open places of the

The parallel vertical keels at the constriction between the mid-lobe and side-lobe of *Acriopsis javanica*. Note also the hooded column and the two forward side-arms.

2mm

lowlands throughout Peninsular Malaysia. The plant clasps onto the trunks of trees. It is hardy and can flower freely.

Acriopsis ridleyi Hk.f.

Plant: has the habit of *A. javanica*
   The following are the differences between the two species:
1) the plant is smaller in size than *A. javanica*
2) the flower structure is similar except that:
   a) in *A. ridleyi* the sepals and petals have dark brown spots
   b) the mid-lobe of the lip is broader, making the constriction between the side-lobe and the mid-lobe more prominent. The apex of the mid-lobe is slightly notched.
3) *A. ridleyi* is less widespread than *A. javanica*. It is found growing on tree trunks in open areas on the hills.

Cultivation:
   Acriopsis is not cultivated by orchid growers. This is possibly because the flower is too small. Few, except the botanist, could appreciate it. The plant flowers rather freely though, and it ought to be an interesting item to have in the garden. Acriopsis is rather hardy. The Pseudobulb forms a thick cluster clasping the tree trunk.

AERIDES

This is a genus of 35 to 40 species. Two are native to Peninsular Malaysia: *A. odorata* and *A. multiflora*. The distribution of the Aerides extends from the Himalayan regions to Indo-China and South East Asia.

The genus was established in 1970 by Loureiro, a Portuguese priest who at that time studied the Indo-Chinese orchids. The type species of the genus is *Aerides odorata* Lour., a species rather prevalent not only in Indo-China but also in the entire region of South East Asia.

The generic name Aerides is derived from two Greek words, *aer* = air, and *eides* = resembling. Together they mean "air plant" or "children of the air", alluding to the epiphytic habit of the plant as "it possesses the power of living almost entirely upon the matters which it absorbs from the atmosphere".

The apex of the column in *Aerides odorata* looks like the beak of a bird.

Aerides odorata Lour.

Previous Names:
Epiderdrum aerides Raeusch
Limodorum latifolium Thumb. ex. Sw
Orxera cornata Rafin.

Plant: is short, grows drooping down.
Leaf: is long with tip slightly bilobed and unequally rounded.
Inflorescence: is pendulous, unbranched with clusters of waxy, white flowers.
Flower: The dorsal sepal and the lateral petals are about the same size. The apex of each is rounded. The lateral sepals are broad and horizontal. The lower margin of the sepal is semi-circular, curving upwards.
The sepal and petal are white with markings of violet-purple. Colouration and pattern vary in different plants. The lip is spurred

10mm

with its apex curving upwards in front. The tip of the spur is greenish or yellowish in colour.

The column is short, the apex is shaped like the beak of a bird. Indeed, at a close look the flower appears like a dove with two wide-spreading wings about to perch on a branch.

Aerides odorata is an extremely variable species. It differs in colouration and minor morphological features depending on the habitat from which it originated. Many authors may have erred by regarding the variants as distinct species when in fact they ought to have been considered as only varieties of the same species, *A. odorata*. In the literature the following variants have been recorded:

var. *suavissimum*. This is considered one of the finest and most scented of the Aerides species.

var. *bicuspidata*. This has a bicuspidated lip.

var. *cornata*. This is a plant from Burma.

var. *virens*. This is a plant from Java.

var. *immaculata*. This is the alba or white form of the species and is found in Northern Thailand.

Two other types, which are considered by the Filipinos as distinct species, are *A. lawrenceae* and *A. quinquevulnerum*. They have larger flowers with brighter and different colour patterns. According to Holttum they do not differ from *A. odorata* in their essential structures and ought to have been called varieties of *A. odorata* instead. Indeed, further study of the plants concerned has to be carried out to verify the exact status of this suggestion.

Aerides odorata is distributed widely from regions of the Himalayas through to Burma, Thailand, Indo-China, Malaysia, Philippines and the Indonesian Islands. In Malaysia, the plant is common on trees growing in moderately exposed areas of the lowlands, especially near the sea.

Cultivation

The Aerides are rather hardy plants and can be grown in pots with loose medium or better still in wooden baskets. They also grow well when tied to a tree trunk or wooden poles. The plant flowers once a year and can throw out numerous flower spikes at a time. When in bloom, the plant is indeed attractive. Give the plant plenty of light and air for good growth.

ANOECTOCHILUS

This is one of the many terrestrial orchids commonly called the Jewel orchid. The plant is delicately succulent and grows amongst the leaf litter under the forest floor. It has beautiful ornate leaves. The white flower, however may not be considered attractive at all although its structure is unique.

There are some 25 species of Anoectochilus distributed from regions of the Himalayas through tropical South East Asia and the Polynesian Islands.

Holttum listed 6 species (Ridley listed 7) which are native to Peninsular Malaysia. Seidenfaden in his study of the genus in Thailand added three more species which are found in Peninsular Malaysia.

1) *A. pomrangianus* Seidenf. - found in Cameron Highlands
2) *A. tonkinensis* Gagnep. - found in Cameron Highlands, Taiping Hills and Gunung Batu Puteh.
3) *A. repens* (Downie) Seidenf. & Smitin. - found in Fraser's Hill.

The generic name is derived from Greek, *anoektos* = open, and *cheilos* = lip. This refers to the spreading blade of the flower giving the appearance of openness.

Anoectochilus albolineatus Par. et Rchb. f.

Plant: is short and the stem succulent with lower portion creeping on the ground. Some short roots develop along the nodes.
Leaf: is ovoid. The apex is short but pointed. It is orangy purple with prominent veins. There are 3 to 4 leaves per shoot.
Inflorescence: is terminal arising from the centre of the leaf-crown. Each inflorescence carries about 3 to 4 flowers.
Flowers: The stalk is brown and hairy with persistent bract. The dorsal sepal forms a hood to cover the column. This sepal is made up of two halves. The upper half is brown, leathery and covered with hairs. The lower half of the sepal is white and membranous. The two lateral sepals are smaller than the dorsal sepal. They are

brown and hairy on the outside. The prominent feature of Anoectochilus is its complicated lip. The mid-lobe is narrow and is grooved, dividing the lip into two equal halves. On each side of the mid-lobe are 10 to 12 fine white claws which bend upwards. At the end of the mid-lobe are two broad-blades. The flower is spurred and this is distinctly bilobed. According to Holttum, *A. albolineatus* is "apparently without pink spots on the lip". This is not so with the species illustrated here. There are two prominent pink spots on each side of the bilobed lip.
The column is attached on top of the spur and is also parallel to it. On the column is an oval cavity in which is lodged the pollinia.

Three species in this genus: *A. albolineatus*, *A. geniculatus* and *A. reinwardtii* are all alike. The feature which can separate *A. albolineatus* from *A. geniculatus* is the spur. In *A. albolineatus* the spur is bilobed, whereas the spur of *A. geniculatus* is not. According to Seidenfaden the Javanese species *A. reinwardtii* is more or less indistinguishable from *A. albolineatus*. The spur in the Javanese species also has a bifid apex. It seems that further studies of fresh samples of the two species may prove that the so called *A. albolineatus* should actually be *A. reinwardtii*. Ridley has recorded that *A. reinwardtii* was also found in Perak and Taiping Hill.

Anoectochilus albolineatus is distributed from Tenasserim down through Thailand and Malaysia. It is found in the wet forest floor in elevated places.

Cultivation

Anoectochilus is a delicate plant and does not grow well in the garden. For this reason it is not popular with collectors. It needs cool, humid and shady conditions for good growth. Use decomposed leaf litter or humus as growing medium.

The flower of *Anoectochilus albolineatus* has hairy sepals with claw-like filaments on its lip.

4mm

APPENDICULA

There are about 40 species of Appendicula distributed throughout the region of the Himalaya down through Malaysia, Philippines, Indonesia and New Caledonia. Of these about 11 species are native to Peninsular Malaysia.

Members of the Appendicula can be small or large plants. They can be found growing on moist rocks or tree trunks. The long, slender stem grows hanging down. One obvious feature about this plant is the leaves arranged in two ranks. The base of each leaf becomes twisted slightly so that the blade comes to lie in one plane. As such the leaf appears flat.

The generic name is derived from the Latin *appendicula* which means small appendage. This is in reference to the characteristic curled appendage found at the base inside the lip.

Appendicula cornuta Bl.

Previous Names:
Appendicula bifaria Lindl.
Appendicula cyclopetala Schltr.

Stem: is long and pendulous.
Leaf: is held at an angle of about 45° to 60° in relation to the stem. Its base is not much twisted.
Inflorescence: is short and is usually at the end of the stem. Each inflorescence has a cluster of about ten flower buds with two or three openings at a time.
Flower: is small. The flower stalk is green in colour and is subtended by a green bract.
The dorsal sepal is white and is hood-like. The sepal and petal are about the same size and are white in colour. The lateral sepal forms the mentum. About half of the lip is broad and bends over. It has faint mauve blotches. Inside, on the lip is a white callus.

The inflorescence of *Appendicula cornuta*. Only one or two flowers appear at a time.

The top of the column is U-shaped. Underside that U is placed the stigma.

A. cornuta can be found in India, Hong Kong, Philippines, Indonesia and Malaysia. Depending on the habitat, the plant can be small or large. It grows in the lowlands or in the areas up to an elevation of 2,000 ft.

Appendicula pendula Bl.

Previous Names:
A. maingayi Hk.f.
A. latibracteata J.J.S.
A. lancifolia Hk.f.

Plant: is pendulous or grows erect with a simple stem.
Leaf: is leathery and broad, arranged in two rows.
Inflorescence: is borne at the end of the stem. It is often branched, thus forming a cluster hanging down. The flowers are well arranged on the inflorescence.

Flower: is subtended by green persistent bract. The bract is about the same length as the flower. The flower bud is green when young but it turns yellow as it blooms. Each flower opens in succession. The flower stalk is longer than the entire flower. The dorsal sepal and lateral petals are about the same size. They are ovoid with pointed apex. The three structures form a hood-like cover over the column. The lateral sepals do not unite but come together to form the mentum. The lip is broad with a round apex. In the middle of the apex is a papillae. The lip forms a hollow channel or throat down the mentum. On the edge of the blade of the lip are low keels, and they continue down the throat where they form a horse-shoe structure.

The column is very short. The anther cap is pointed and appears like the beak of a bird.

Appendicula pendula is distributed in Indonesia and Malaysia. In this country it is fairly common in both the lowlands and mountains. Because of this, the size of the plant and its flower could vary.

|— 2mm —|

The horse-shoe-like structure situated down the throat of the lip in *Appendicula pendula*.

Each flower is subtended by a green persistent bract. The flower is green at bud stage but turns yellow when open.

|— 3mm —|

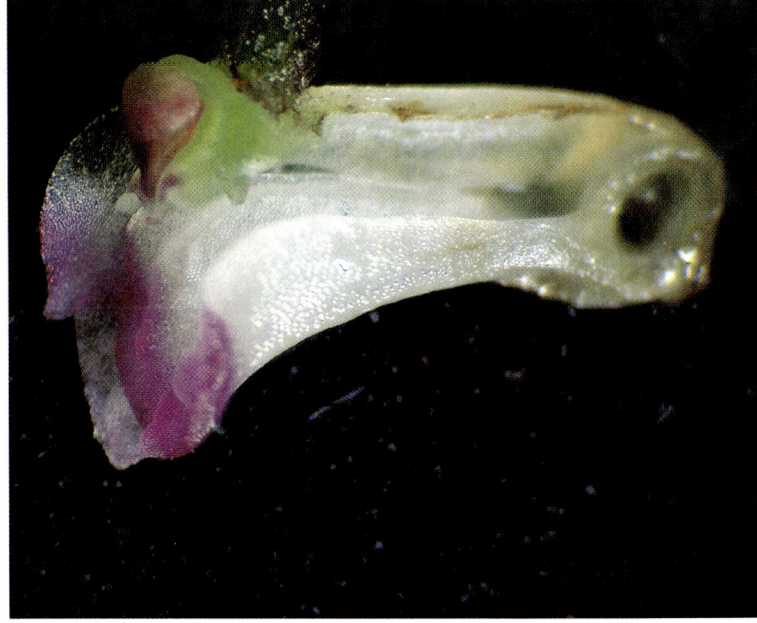

Appendicula undulata showing the curled appendage at the base of the lip.

1mm

Appendicula undulata Bl.

Previous Names:
Appendicula purpurascens de Vr.
Podochilus unciferus Hk.f.

Stem: is pendulous and unbranched.
Leaf: is small, broadly elliptic and is distichously arranged to give a flat appearance.
Inflorescence: is very slender, with 1 or 2 flowers near its tip. It is covered with several green sheaths.
Flower: The dorsal sepal is oval but tapers to a point. It is white with a faint mauve line in the middle. The lateral sepal is white and forms the mentum. The petal is short, oval and is bright purple towards the apex. The lip is long. It is white except towards the tip which is bright purple. At the base of the lip is a curled appendage. The column is short with two side arms protruding out like horns on the head. It is green in colour.

The short column with two protruding side-arms in *Appendicula undulata*.

A. *undulata* is found in the Philippines, Indonesia and both East and West Malaysia. The plant grows as an epiphyte on tree trunks, or on boulders along streams of montane areas from 2,000 to 5,000 ft. elevation.

Cultivation

Appendicula is not cultivated by orchid hobbyists. The genus however, makes an interesting botanical specimen. It grows well in shady areas attached to a wooden support. The plant is free flowering in cultivation.

1mm

ASCOCENTRUM

This genus can be referred to as the Miniature Vanda. It has a similar growth habit as the Vanda, and its small flower, if enlarged resembles the Vanda flower.

The Ascocentrum is essentially a genus belonging to South East Asia and has only a few species. Thailand has three very popular species while in Malaysia, there is only *A. miniatum*.

The generic name is derived from the Greek *ascos* meaning bag, and *kentron* meaning spur. The name possibly refers to the prominent spur which hangs in front of the lip.

Ascocentrum miniatum (Lindl.) Schltr.

Previous name:
Saccolabium miniatum Lindl.

Plant: is diminutive, hence the specific name *miniatum*.
Leaf: is fleshy. Its upper surface is channelled. The apex is lobed and toothed.
Inflorescence: is erect with a cluster of well arranged flowers.
Flower: The stalk is long, about one and a half times the length of the spur. The sepal and petal are about the same size. They are bright orange or orange-yellow. The apex is rounded. The lip consists of the mid-lobe which is shaped like an elongated tongue. The side-lobe consists of a thick horizontal outgrowth covering the opening of the spur. The spur is laterally compressed.
The column is short and stout.

Ascocentrum miniatum is distributed in regions extending from the Himalayas to Java. In Peninsular Malaysia, the species is only found in the northern states. The plant is found growing on trees in exposed places.

Cultivation

This species is popular with orchid growers for the flower is attractive. In Thailand, commercial growers selected the good plants and hybridized them. The resulting hybrids are sold to hobbyists. A plant about the height of 3 cm can start to bloom. Ascocentrum can be grown tied to branches or the trunks of trees or grown in hanging wooden baskets. It can also grow well in a clay pot provided with broken bricks or charcoal.

Ascocentrum miniatum, a native of the northern states of Peninsular Malaysia.

ARUNDINA

Arundina is a genus of terrestrial orchids. In Peninsular Malaysia it is seen growing in clusters in open spaces along roads. At first glance, Arundina resembles tall grass, hence its common name, the Bamboo orchid.

The generic name is derived from Latin, *arundo* meaning reed. Obviously this is in reference to the resemblance of the stem to a reed.

Arundina graminifolia (D. Don) Hochr.

Previous names:
Bletia graminifolia Don.
Arundina bambusifolia Lindl.
A. speciosa Bl.
A. chinensis Bl.
A. densiflora Hk.f.
A. affinis Griff.

Plant: grows in clusters. The stem is slender and erect.
Leaf: is narrow and grass-like.
Inflorescence: is terminal, sometimes it is branched. It elongates gradually, producing a succession of flowers opening one or two at a time.
Flower: Each flower is subtended by a small, stiff bract. The sepal is narrow and pointed while the petals are broad. At first glance the flower resembles the Cattleya. The lip is broad and not lobed. The outer edge is frilled. Running through the middle of the lip are four ridges or keels of which the middle two are prominent and longer. There are some yellow blotches towards the apex of the lip.
The column is about half the length of the lip. It is white, broad and winged.
The colour of the flower is variable and the following flower types can be distinguished:

The cattleya-like flower of *Arundina graminifolia*, the Bamboo Orchid.

1) Arundina with large, rosy mauve flower
2) Arundina with small, rosy mauve flower
3) Arundina with white flower

Arundina graminifolia is widely distributed, extending from Sri Lanka to the Himalayan region, South China, Indo-China, Thailand, Malaysia and down to the islands of the Pacific. Surprisingly enough the genus is not represented in the Philippines.

Several species of Arundina have been described by various authors, but according to Holttum, they are but variants of the same species.

The Himalayan Arundina, first known in 1825 as *Bletia graminifolia,* is about 6 feet tall. The flower is large and has no yellow on the lip. A dwarf variety known as *A. minor* is native to the montane areas of Sri Lanka, and was named *A. minor.* There is also a white Arundina from Sumatra known as *A. sanderana.* Even in Peninsular Malaysia, plants collected from different localities show differences in characteristics such as height, size and spread of leaves, branching size, shape and colour of flower parts, etc. It has been written that "no two specimens are exactly alike".

In Peninsular Malaysia, Arundina is found in the lowlands as well as in the mountains. It inhabits open, sunny places - never the shade of the forest.

Cultivation

Arundina graminifolia or the Bamboo orchid can be a good plant for the garden. Sometimes it has been grown in the garden by people who never even realize that it is an orchid. It can be grown in pots or as a bedding plant. Since it is not free-flowering, a number of plants need to be grown together to be anything pretty or worthwhile. The Bamboo orchid should be grown in ground provided with well drained friable soil. It should be exposed to full sunshine. When planting, care should be taken not to place it as deep in the ground as to cover the new shoots. This may kill the young shoots.

The flower of the Bamboo orchid does not last long. It stays open for 3 to 4 days, after which it fades away. After the stem has finished flowering, it may develop shoots at its nodes. These shoots can then be separated and rooted for new planting.

BULBOPHYLLUM

This is the largest genus in the Orchid family. According to conservative estimates there are at least 1000 species in this genus. Alex Hawkes in his book *Encyclopedia of Cultivated Orchids,* regarded Bulbophyllum as a 2000 species genus.

Bulbophyllum is distributed worldwide, from tropical America to Africa, Asia, Japan, the Pacific Islands down to as far as New Zealand. New Guinea has at least 569 species and this is the largest number of species found in any one area. Holttum listed about 126 native species of Peninsular Malaysia. Brazil has about 100 species.

Due to the large number of species, generic characteristics are not clearly distinguished. As such the genus has been divided into sections or sub-genera by various people. Some authors even raised the status of certain sections to that of genera. One well-known example is the "fairly widely accepted genus", Cirrhopetalum, established by Lindley in 1924. This group is separated from the rest of the Bulbophyllum on account of the fact that the inflorescence has an umbel-like arrangement. The name Cirrhopetalum comes from the Greek *kirrhos* which means yellowish, and *petalon* or leaf. This is in reference to the prevailing yellow colour which Lindley always found to be present on the sepals of the flowers he had seen. Since then, Cirrhopetalum has been a "football". Later, Reichenbach put it back into the genus Bulbophyllum, but later changed his mind and reinstated it to its original status. Schlechter also recognised it as a genus. J.J. Smith favoured a separation of the two. From this, it appears that a thorough study of the genus is needed to clear any mispresentation or misinformation that may exist. Holttum admitted that the account of Bulbophyllum as presented in his book was imperfect. Indeed no author could claim unchallengeable knowledge of this genus. Perhaps it might be too late now to suggest a detailed re-examination of our native species. It is most likely that they may have been destroyed and lost forever!

The generic name Bulbophyllum is derived from two words, *bolbos* = bulb, and *phyllon* = leaf. This refers to the leafy pseudobulb. The genus was established by Thouars in 1822, using a specimen from Africa as a type-species. Botanically, Bulbophyllum is related to Dendrobium and Eria. In fact, they seem to share a common habitat. It is possible to find these three types of orchids growing on the same host.

Bulbophyllum has two distinctive characteristics:
1) The pseudobulbs are conspicuous, angular and single leafed. They are joined to each other by a short or long rhizome. The pseudobulbs vary in size. They can be erect or prostrate on the rhizome. So, vegetatively, Bulbophyllum is distinctive by itself. One other group of orchids which may resemble the Bulbophyllum vegetatively is the Dendrochilum.
2) The semi-circular lip of the flower is succulent and characteristic of the genus. It is hinged to the column-foot and can swing like a pendulum. This is indeed a marvellous and interesting adaptation for pollination. An insect visiting the flower lands on the lip which then tilts back and forth. By this action the back of the insect rubs against the column above, resulting in the possible removal of the pollinia which then adheres onto its back, or if the pollinia is already on the insect's back, it will then adhere to the sticky stigma.

Bulbophyllum biflorum T. et B.

Previous names:
Cirrhopetalum biflorum J.J.S.
Bulbophyllum geminatum Carr.

Pseudobulbs: have 4 angles and are quite apart from each other. The leaf is narrow and oblong with a blunt apex.
Inflorescence: is long, at the end of which are two flowers. The top of the flowers are close together but their "feet" are apart.
Flower: The dorsal sepal is short and forms a hood over the column. It is purple brown. The apex is pointed. The lateral sepal is narrow and long, forming a thin tail which terminates in a thick-ended tip. The sepal is smaller than the dorsal sepal and is broadest in the middle. It tapers to a narrow flat apex ending with a short frill. The lip is thick, curved, and is hinged to the curved column-foot.

B. biflorum is quite a rare species. It can also be found in Indonesia.

The two flowers of *Bulbophyllum biflorum*.

Bulbophyllum corolliferum J.J.S.

Previous names:
Cirrhopetalum curtissii Hk.f.
Cirrhopetalum concinnum Hk.f.
Bulbophyllum pulchellum var. *purpureum* Ridl.
Bulbophyllum curtissii J.J.S.

Pseudobulbs: are ovoid and angular. They are separated from each other by a short rhizome. The leaf is succulent, broad and long. The apex is blunt and slightly bilobed.
Inflorescence: is long and slender. There are 8 to 12 flowers borne in a whorl at the end of the inflorescence.
Flower: The dorsal sepal is oval, concave and forms a hood over the entire flower. The apex is fine and long. It is entirely purple. The two lateral sepals are broad and touching one another at the front. About half the length of the basal end of the sepal is flushed with pale yellow.
The petal is about the size of the dorsal sepal but the shape differs. It is oval, about $^3/_4$ its length broad and tapering towards the apex. The edge has frills. The petal is brown with some yellow markings at the base.
The column is short and is winged on the underside. It is yellow with some purple dots or bars sparsely and randomly distributed throughout. The column-foot is curved to form a semi-circle and onto it is hinged the succulent, curved purple lip.

B. corolliferum is found throughout the lowlands and highlands of Peninsular Malaysia. Its distribution also extends to Thailand. It is commonly found on trees.

There is a discrepancy with regard to Holttum's description of this species (Flora of Malaya: Orchids pg: 414). In his book, Holttum described the flower of *B. corolliferum* as "entirely purple except the tip". The specimen illustrated here has yellow flushes at the base of the sepals.

Bulbophyllum corolliferum with flowers borne in a whorl.

9mm

Bulbophyllum lilacinum Ridl.

Rhizome: is about half-pencil size, brownish and wrinkled.
Pseudobulbs: are about 4 to 6 cm apart. They are green, narrowly ovoid and appear succulent and smooth.
Leaf: is fleshy and oblong with a blunt apex.
Inflorescence: is pendulous and is made up of a compactly arranged reddish cluster of flowers. The inflorescence stalk is reddish brown with specks of green.
Flower: is subtended by a long persistent bract which is light brown in colour. The bract is about two times the length of the flower stalk. The flower stalk is about the same length as the dorsal sepal. It is light green with streaks of purple. All the sepals and petals are similarly coloured.
The dorsal sepal is a pointed hood-like structure. The two lateral

sepals are broad and taper forward to meet one another. The lateral sepals are reduced to form short, narrow, pointed structures. The lip is pale yellow and curved. From the front it appears like a tongue. On each side of the tongue is a small pointed lobe. The column is short with a hollow, in which the pollinia are slotted in. This hollow has three horns or pointed structures.

B. lilacinum is found in Peninsular Thailand and the northern states of Peninsular Malaysia.

Bulbophyllum lobbii Lindl.

Previous name:
B. siamense

Rhizome: is hairy and wiry, producing roots between the nodes.
Pseudobulb: has one thick and fleshy leaf, shaped like a spatula.
Inflorescence: is single flowered. The flower stalk is long and stiff.
Flower: does not seem to open fully, i.e. is not spreading. The sepal is larger and broader than the petal. It is pale yellow with reddish purple mottling. The upper sepal is erect and tapers towards a point. The basal margin of the lateral sepal is semi-circular with its apex curving downwards. The petal is narrow and is rather pointed. The lip does not "swing" as in most of the Bulbophyllums. It is short and broad, shaped as in a classical drawing of the heart, i.e. the apex

The flower of *Bulbophyllum lobbii* of Peninsular Malaysia is different from that found in East Malaysia.

narrows to a pointed end. On the opposite end it is broad with a wide, gentle notch. At the base of this notch is a large blotch of orange yellow while the rest of the lip has numerous fine purple dots.

The column is broad and short. It is deep orange yellow in colour. The column-foot has deep purple streaks.

Bulbophyllum lobbii is a common mountain plant. The species found in Peninsular Malaysia has flowers which are much different from those of the Borneo or East Malaysian plant.

Bulbophyllum longiflorum Ridl.

Rhizome: is wiry
Pseudobulb: is about 4–5 cm apart. It is slightly tapering.
Leaf: is leathery, narrows slightly upwards and is widest above the middle. The apex is acute.
Flower: is solitary. The flower stalk is slightly longer than the petal. The dorsal sepal is long and tapers to a tail. The lateral sepals are close together. The lateral petals are about horizontal and at the tips curve down forming a long "moustache". Both the sepal and petal are pink with darker veins. The lip is deep orange yellow. It is curved with a blunt, cleft apex. There is a median line running along the lip as if dividing it into two equal halves. On each side of this line is a blotch of purple. The underside of the lip is rather flat and there are numerous yellow bristles. The lip is attached to a curved column-foot. The upper side of the column-foot is purple. The column is short and has two pointed arms on the underside.

Bulbophyllum longiflorum, a plant in need of a new name.

A close-up of *Bulbophyllum longiflorum* showing the short column with the arms on its underside. Note also the numerous yellow bristles under the lip.

3mm

This species needs to be given a new name and is not in any way related to another one which is also know as *B. longiflorum* Thou. When Ridley gave the name to this species he must have not been aware that Thouars had made use of this name already. Seidenfaden, suggested to the author:"if you are bold enough to propose a new name, be sure to look up at least all the names in Index Kewensis first".

B. longiflorum is found among rocks and boulders of montane areas of northern Peninsular Malaysia.

Bulbophyllum maximum (Ridl.) Ridl.

Previous name:
Cirrhopetalum maximum Ridl.

Bulbophyllum maximum are arranged in a ringed cluster.

Detail of column with a bristle on each horn.

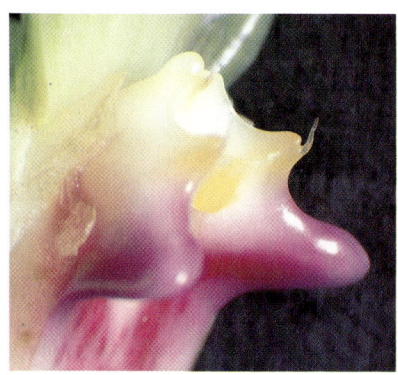

3mm

The lip of *Bulbophyllum maximum* swings like a delicate pendulum.

|—— 4mm ——|

Rhizome: is long and thick and covered with a sheath.
Pseudobulbs: are narrow and oblong and about 20 cm apart from each other. The leaf is lanceolate, leathery, and has a petiole.
Inflorescence: is single, bearing about 6 to 8 flowers arranged in a whorl. As a whole, the inflorescence appears like a drooping straw hat.
Flower: is foul-smelling. The sepal is broad and narrows down to a pointed tail. The mid-vein is prominent. The lateral sepal is joined to the underside of the column-foot. It is slightly longer than the dorsal sepal. The sepal is much smaller than the petal and is about $1/4$ its length. The lip is succulent, curved and broad. It tapers to a point. The centre of the lip is purple and the edge white. The column is short and is purple in colour. On the underside of the column are two protruding arms. The three corners of the pollinia cavity are raised and horn-like, each with a short bristle.

B. maximum can be found growing on rocks by streams or on trees. It grows in the lowlands up to a moderate elevation. The species is also found in Indonesia.

Bulbophyllum medusae (Lindl.) Rchb.f.

Previous name:
Cirrhopetalum medusae Lindl.

The plant is similar to *B. vaginatum* both vegetatively and in the shape of the flower. The following are features which can distinguish *B. medusae* from *B. vaginatum* (Lindl.) Rchb.f.
1) The lateral sepal of the former is much longer and is white with some pink spots.
2) The edges of the sepal and petal of the former do not have 'teeth'.

B. medusae is not a common species. It grows on rocks and trees. During flowering season, the plants sometimes burst into massive blooms. They grow all over Peninsular Malaysia besides Thailand and East Malaysia.

Bulbophyllum medusae, growing in a cluster and flowering at the same time.

Bulbophyllum patens King

Rhizome: is thick and covered with a fibrous sheath.
Pseudobulbs: are small, with leathery, broad spatula-shaped leaf. Each pseudobulb is rather far apart from the next.
Flower: is single and originates from the node of the rhizome. The flower gives an appearance of a scorpion's head and it is orientated lip inwards. So, what one sees is the back of the flower. The flower stalk is long and yellowish green with numerous spots. The dorsal sepal is joined to the column-foot and is horizontal before curving upwards to give a broad "U" shape. The petal is about $3/4$ length of the dorsal sepal. The lip is fleshy and is curved. The entire flower is yellow with numerous purple spots throughout.

B. patens is found on trees throughout the lowlands of the country. The species is also found in Indonesia.

Bulbophyllum sessile (Koen) J.J.S.

Previous names:
Epidendrum sessile Koen
Bulbophyllum clandestinum Lindl.

Rhizome: is slender and creeping. It is much branched and grows hanging down. The rhizome is covered with a brown sheath.
Pseudobulbs: are small, green and appressed to the rhizome.
Leaf: is green and fleshy. One leaf grows out from the end of each pseudobulb. The apex of the leaf is blunt but with a tiny tip.
Flower: is cream yellow borne singly along the nodes of the rhizome. The flower stalk is short. The sepal's width is about $1/2$ its length. It tapers to a long, narrow end. The three sepals cover the

entire flower. The petal is as long as the lip and is about $1/2$ the length of the sepal. The lip is not lobed and is tapered. The column is short.

B. sessile is a common epiphyte found on wayside trees, orchards, etc. It often grows as a big mass. It is found throughout Peninsular Malaysia. The distribution of this species also extends to Indonesia, New Guinea and northwards to Indochina.

Cultivation

Bulbophyllum, in spite of their large numbers, are generally not cultivated. Only few species are worth mentioning. *B. maximum* is an attractive species and appears to be worth cultivating, although it is foul-smelling. This possibly keeps the plant off most collectors' gardens. *B. corolliferum* and related species are worth cultivating. The plant is small and makes a cute specimen. The flowers, though small, are attractive and deeply coloured. Bulbophyllum needs a shady place and fairly moist conditions. It should be tied to a wooden support or allowed to scramble over large rocks.

CHILOSCHISTA

The genus Chiloschista is distributed from India, Thailand and the Malaysian regions down to North Queensland. There is, however, no mention of this genus in both the books written by Ridley and Holttum, implying that no Chiloschista has ever been recorded in Peninsular Malaysia. In 1966, Professor Holttum corrected this omission. A Chiloschista was found by Mr. Lim Swee Lim in northern Perak and this plant was sent to Holttum at Kew through Mr. Cheang Kok Choy, then Head of the Penang Botanical Gardens. With this, Holttum made the first description of the plant and called it *Chiloschista sweelimii,* obviously in honour of the finder of this plant.

The name Chiloschista originated from two Greek words, *cheilos* = lip, and *schistos* = cleft. This refers to the cleft lip of the flower. There are about 3 to 4 species of Chiloschista. They are unique in that the leaf is very small or much reduced and is not obvious. What can be seen of the plant is a cluster of green roots. These roots carry out photosynthesis, usually the function of the leaf in other plants. In *C. usneiodes* the root is flat and was thought to be a leaf.

Chiloschista sweelimii Holtt.

Leaf: is minute, about 4 to 5 mm. It is not green and withers away.
Root: is cylindrical and green.
Inflorescence: is long with 10 to 20 flowers. Each flower is subtended by a small thin hairy bract. The sepal is covered with erect hair on both surfaces. The hairs on the back surface are longer. The sepal and petal are pale yellow with orange spots throughout. The lip is about 3 mm in length. The mid-lobe (about 1 mm) is fleshy, rounded and curves downward. It is pale yellow. The side-lobe tapers towards the lip. Its edge adjacent to the mid-lobe is almost straight and erect but the edge near the column is curved. Exter-

Chiloschista sweelimii, a species found by Lim Swee Lim in Northern Perak. This is the first Malaysian Chiloschista species described by Prof. Holttum in 1966.

nally it is white but internally has red-brown bands. The spur is pale yellow with small orange spots. Internally, it is covered with white, club-shaped hair.

The column is about $1^1/_2$ mm long and is pale green.

C. sweelimii is not common and was recorded in the forest areas of northern Perak or central Peninsular Malaysia.

Cultivation

This is a unique plant worth cultivating. The roots grow in clusters on twigs. If well grown numerous arching inflorescences arise and each bears attractive flowers that can last for 2 to 3 weeks. The plant should be grown on twigs or wood pieces under partial shade.

CLEISOSTOMA

The genus Cleisostoma was established by Blume in 1825. The generic name is derived from the Greek *kleistos* = closed, and *stoma* = mouth. This genus refers to a group of plants that has a callus or plug at the entrance of the spur.

There has been some confusion in assigning members to this genus. Ridley, in the book *Flora of the Malaya Peninsula: Vol. 4* was unable to separate the genera Cleisostoma and Sarcanthus based on the presence of a callus in the spur. Therefore, he lumped all species of the two "genera" into one large genus called Saccolabium. Saccolabium is Greek for "bag-like lip" (*saccus* = bag, *labium* = lip). Later Holttum classified what are now considered species of Cleisostoma into the genus Sarcanthus. The generic name Sarcanthus means fleshy, referring to the succulent texture of the flower. After studying the details of the plants under this genus, it seems that "fleshy" is not an appropriate description. Compared to other genera, such as Pomatocalpa, the Sarcanthus flower is not that "fleshy" after all. Besides, Garay pointed out that the use of Sarcanthus as a generic name is unjustified. So 14 out of the 16 species listed by Holttum as Malayan Sarcanthus should rightly be known as Cleisostoma. In doing so, it becomes consistent with the description of the genus, that is, that the flower has a callus at the mouth of the spur.

Cleisostoma is distributed throughout South East Asia, Australia and the adjacent islands. The genus consists of about 100 species.

Cleisostoma discolor Lindl.

Previous names:
Sarcanthus termissus Rchb.f.
S. josephii J.J.S.

Stem: is short and stout.
Leaf: is not thick. The apex is rather deeply and unequally bilobed. There are about 10 leaves to plant.
Inflorescence: is pendulous, simple or can be branched. Each inflorescence has many, compactly arranged flowers.
Flower: The flower stalk is long. The sepal is slightly narrower than the petal. It is yellowish green with a median purplish-red band. Both the petals are curved forward. The lip is prominently lobed. The side-lobes are erect and are curved to form two semicircles touching each other. The edge is whitish while the base has reddish purple patches. The mid-lobe is concave, but its apical half bends downwards to form a tongue. The edge of the tongue is wavy. It is whitish with spots of purple. The spur is slightly longer than the petal and is orientated about 45° to the flower stalk. It is narrow, tapering to a blunt apex. The back of the spur is vertically grooved dividing it into two equal halves. At the mouth of the spur is a callus. This originates from the back wall of the spur and appears to be an extension of the column-foot. Within the spur is a thin membrane dividing the cavity into two vertical halves.
The column is short. The anther-cap has a prominent broad median line of purple.

C. discolor is found in the northern states of Peninsular Malaysia. Its distribution extends to the neighbouring Indonesian islands.

Cleisostoma discolor, a species previously known as *Sarcanthus termissus*.

Cleisostoma scortechinii (Hk.f.) Garay

Previous names:
Sarcanthus scortechinii Hk.f.
Saccolabium scortechinii Ridl.

Stem: is short and grows hanging down.
Leaf: is leathery and seems to be held in one plane, thus the plant appears flat.
Inflorescence: is single, and hangs down.
Flower: is rather small The sepal is broader than the petal. The dorsal sepal is broadly oval while the lateral sepal is slightly rectangular with a skewed rounded apex. In the middle is a pale yellow line. The petal is narrow. Both the sepal and petal are reddish-brown in colour. The side-lobe is erect and windy. The mid-lobe is shaped like a broad arrow-head. Its apex is slightly cleft in the middle. This arrow-head mid-lobe distinguishes this species from *C. subulatum*. At the base of the lip are two calli which fit closely into another callus at the mouth of the spur. The conspicuous white spur is short and tapers to a blunt apex. It is orientated about parallel to the flower stalk. The callus at the mouth of the spur originates from the back wall of the spur. It is yellow, round and large. Opposite it, the wall of the spur is thick and succulent. A vertical membrane divides the cavity of the spur into two halves.

C. scortechinii is found throughout the lowlands of Peninsular Malaysia. It grows on trees by the sea or on branches overhanging rivers. The species is widely distributed in the area ranging from India, Thailand, Indo-China through to the Indonesian islands.

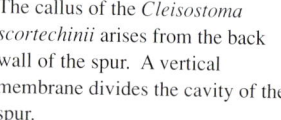

The callus of the *Cleisostoma scortechinii* arises from the back wall of the spur. A vertical membrane divides the cavity of the spur.

1mm

Cleisostoma scortechinii showing its arrowhead-shaped lip.

5mm

Cleisostoma subulatum Bl.

Previous names:
Sarcanthus subulatus (Bl.) Rchb.f.
Sarcanthus secundus Griff.
Saccolabium secundum Ridl.

Cleisostoma subulatum, note the two lumps on the column.

Stem: grows hanging down.
Leaf: is long and fleshy and the tip is sharply pointed. The leaf is thus different from *C. scortechinii* where it is flat and broad.
Inflorescence: is long and pendulous with numerous flowers held closely together. The colour of the newly opened flowers are faint but with time the colour becomes heavier.
Flower: For each flower, the pointed pinkish spur is most prominent. The flower stalk is about the length of the spur. The sepal is broader and longer than the petal. It is dull brownish-red with greenish yellow median and edges. The spur is cylindrical and tapers to a blunt end. The upper part of the spur together with some parts of the mid-lobe is tinted mauve. The lower half of the spur is white. The side-lobe is erect and is yellow in colour. The mid-lobe tapers and is "V" shaped. It is tinted violet but the tip is white.
Arising from the inner wall, at the mouth of the spur, is a large, prominent two-ridged callus. On the outer wall of the spur arises a membranous structure which divides the spur into two vertical halves but this does not completely partition the chamber.
The column is short and broad. Underside, it forms two lumps in between which is the hollow of the stigma.

C. subulatum is widely distributed and can be found in India, Thailand and Indo-China. The distribution extends southwards through Malaysia to the Celebes. In Peninsular Malaysia the species grows in the lowlands, especially on trees by the sea or on those overhanging rivers.

Cultivation
Cleisostoma is not cultivated by hobbyists though it is easy to grow under lowland conditions. The plant should be tied to a wooden post. Thus, it should grow without difficulty.

A section through the spur of *Cleisostoma subulatum* showing the callus at the mouth of the spur. A membranous structure divides the spur into half vertically.

1.3mm

CLEISOMERIA

This genus which includes two species, was established by Lindley in 1855. The floral structure is similar to the Cleisostoma (or illegitimately Sarcanthus), hence this new name Cleisomeria is given to the genus. This genus has been completely overlooked by the compilers of Index Kewensis and as a result there was some confusion in the classification of this genus. Some placed the species of this genus in Saccolabium while other authors called it Sarcanthus or Cleisostoma.

Cleisomeria lanatum (Lindl.) Lindl.

Previous names:
Cleisostoma lanatum Lindl.
Sarcanthus lanatus (Lindl.) Holtt.
Sarcanthus bracteatus Ridl.
Saccolabium lanatum Hk.f.

Stem: is erect and rather stout.
Leaves: are arranged closely and overlap at the base. They curve like the leaves of a Vanda. They are fleshy and unequally bilobed with rounded tips.
Inflorescence: is branched and pendulous.
Flower: Each flower is subtended by a dark brown bract. Each bract is about $^3/_4$ the length of the flower. The flower stalk is thick in relation to the flower. It is woolly as in the Eria. Flowers are held close together in a cluster. The sepals are stiff and thick. They are arranged in a triangle. The dorsal sepal is lanceolate and hood-like. The outer surface of the sepal is covered with woolly hairs. It is dark brown with a yellow border. The lateral petal is thin with a rounded apex. It is greenish yellow with blotched of brown in the centre. The spur is short and the end rounded. It is held about parallel to the flower stalk. The side-lobe is triangular, and pointing forward. It is pale yellow. The mid-lobe is succulent, white with purple spots. The apex is forked with two pointed arms curving up. At the mouth of the spur is a large callus which originates from the

front wall of the spur.
The column is short. The cavity for the pollinia is deep. The underside of the column has thin appendages.

The distribution of *Cleisomeria lanatum* extends from Tenasserim down through Thailand and Peninsular Malaysia.

Cultivation
The plant is not cultivated. It grows well tied to a wooden post in the shade.

The branched and pendulous inflorescence of *Cleisomeria lanatum*.

CORYMBORKIS

Corymborkis is a genus of about 18 species. It is distributed throughout the shady forest regions stretching from Africa to the Samoan Islands. Three species are found in Peninsular Malaysia: *C. veratrifolia, C. rhytidocarpa* and *C. brevistylis*.

Corymborkis is barely known or studied, much less cultivated. When the author first encountered *C. veratrifolia,* he could not tell that it was an orchid. The plant is unbranched. The leaves are broad, plicate and spirally arranged. This gives the appearance of a small palm or a ginger plant. Indeed, in nature *C. veratrifolia* is found growing among gingers and other herbaceous plants of the forest floor.

The name of the genus is derived from two Greek words, *korymbos* = cluster, *orchis* = orchid, referring to the cluster of white flowers borne at the axil of the leaf.

Corymborkis veratrifolia (Reinw.) Bl.

Previous names:
Hysteria veratrifolia Reinw.
Corymbis veratrifolia (Reinw.) Rchb.f.

Plant: can be tall and unbranched. At first glance it appears as a palm.
Leaf: is plicate, broad and spirally arranged on the stem.
Inflorescence: has 4 to 6 branches with about 2–6 flowers on each branch.
Flower: The sepal is long and narrow giving the appearance of an elongated teaspoon. The apex is acute. The outer side of the sepal is pale green. The two lateral petals are white in colour and are about the same shape and length as the sepal. The lip is as long and rolls over about $^3/_4$ circle to cover the column. Towards the end, the lip becomes broad and has crisped edges.
The column is cylindrical and long. It appears like the head of the spear. The anther is brown, narrow and long.

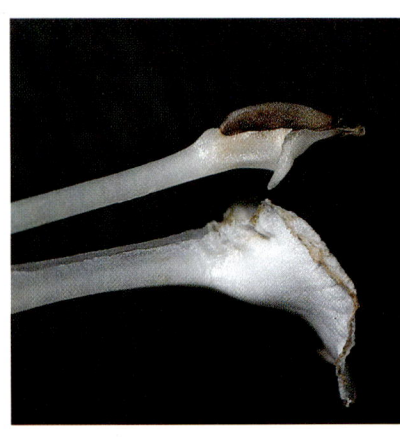

The lip and spear-head column of *Corymborkis veratrifolia*.

0.7mm

The inflorescence of the palm-like *Corymborkis veratrifolia*.

Corymborkis veratrifolia is found in the southern islands of Japan down through the Philippines, Thailand, Malaysia, Indonesia and New Guinea. In Malaysia, the plant is found both in the lowlands and the highlands under the warm, humid shade of the forest.

Cultivation

The plant is not cultivated by orchidists. However, many who have seen it, feel that it is a plant worth cultivating. The flowers come in beautiful white clusters. The plant needs shade and being a terrestrial, does well in ordinary garden soil. Given the right conditions, the plant should bloom once or twice a year. The blooms can last for 1 to 2 weeks. Corymborkis is a fairly easy plant to take care of.

CYMBIDIUM

This is a genus of some 50 species of rather hardy plants. Its distribution extends from Madagascar through to Sri Lanka, India, South East Asia, up north to Japan, and down south to Australia. About 10 species of Cymbidium are native to Peninsular Malaysia. Basically the Cymbidium can be divided into 3 groups. (1) The large-flowered terrestrial species. (2) The miniature terrestrial species. (3) The epiphytic species having long, pendulous flower spikes.

The genus was found by the Swedish botanist, Swartz in 1800. The generic name is derived from the Greek word *kymbes* meaning boat. This refers to the boat-like shape of the lip.

Cymbidium atropurpureum (Lindl.) Rolfe.

Previous name:
Cymbidium pendulum var. *atropurpureum* Lindl.

This plant has similar features as *Cymbidium finlaysonianum*. Its specific name *atropurpureum* refers to the dark purple colour of the flower.

In spite of the similarities, the following are characteristics which distinguish the two species.

	C. atropurpureum	*C. finlaysonianum*
1.	Flower is smaller in size.	
2.	Side-lobe of the lip is shorter than the column.	Side-lobe tapers forward. It is a little longer than the column
3.	Keel is curved.	Keel is straight.
4.	Mid-lobe is longer and a little wider with several irregular purple spots.	Mid-lobe is yellow-white in the middle with a broad crescent-shaped purple patch towards the end.

In *Cymbidium finlaysonianum* the side lobe is longer than the column.

In *Cymbidium atropurpureum* the side lobe is shorter than the column.

Cymbidium atropurpureum is also widely distributed and an epiphyte.

Cymbidium chloranthum Lindl.

Previous name:
Cymbidium sanguinolentum T.et B.

Plant: is a terrestrial and is small in size. The pseudobulb has about 6 leaves.
Leaf: is thin with a rounded tip. It is flat at the apex but is folded along the midrib.
Inflorescence: is erect with many flowers.
Flower: The stalk is about twice the length of the sepal. The sepal is broader and is slightly longer than the petal. It is narrow and green in colour. The apex is blunt. The lip is about twice the length of the column. The side-lobe is erect and rounded. The mid-lobe is nearly circular with a wavy edge. Its centre is green, while the edge is white with purple spots.
The column is pale green with spots of purple.

Cymbidium chloranthum is found in Indonesia, and Borneo besides Peninsular Malaysia.

Cymbidium chloranthum, showing the long flower stalks and the rounded apices of the flower.

Cymbidium dayanum Rchb.f.

Previous names:
Cymbidium acutum Ridl.
C. simonsianum King & Pantl.
C. eburneum var. *dayanum* Hk.f.

Plant: is small and is more of a terrestrial than an epiphyte.
Leaf: is like grass with a pointed tip.
Inflorescence: is more or less erect with 6 to 8 attractive flowers.
Flower: The stalk is about the same length as the sepal. The sepal is about the same size but is slightly longer than the petal. It is narrow with a pointed apex. It is white or mauve with a purple band in the middle. The lobed lip is about the same length as the petal. The side-lobe is slightly longer than the column. It is upright, pointing forward and has a rounded apex. It is pale yellow with purple veins. The mid-lobe is purple with yellow at the basal area. Its lip curves downwards. At the inner half of the basal area are two parallel keels which can be white or yellow.
The column curves slightly and is winged. It is deep purple to about black in colour.

The distribution of *Cymbidium dayanum* extends from Assam in India through to Thailand, Malaysia, Philippines and Indonesia. It was Reichenbach who first described the species in 1869 from the material collected by J. Day from Assam.
In Peninsular Malaysia, *C. dayanum* inhabits the elevated montane area.

Cymbidium finlaysonianum Lindl.

Previous name:
Cymbidium pendulum Bl.

Plant: is an epiphyte with a pseudobulb bearing about 5 leaves.
Leaf: is thick and fleshy. Its apex is rounded and cleft. Each leaf is divided into two unequal halves.
Inflorescence: is pendulous with many well-spaced flowers. It originates from the base of the pseudobulb.
Flower: The sepal and petal are about equal in size. They are greenish yellow flushed with dull purple down the centre. The lip is three-lobed. The side-lobe is erect and comes close to the side of the column. It tapers forward and its end extends a little longer than the column.
The column is about half or three-fourths the length of the lip.

C. finlaysonianum is widely distributed throughout the South East Asian region. It was first collected by Finlayson in Indo-China during the early years of the 19th century. In Peninsular Malaysia it is found throughout the country. It is an epiphyte growing in large clusters on trees in exposed places, especially near the sea.

Cymbidium lancifolium Hook.

Plant: is a terrestrial with the pseudobulb carrying 3 to 4 leaves.
Leaf: is stalked. The leaf blade is rather tough but thin.
Inflorescence: is erect with about 5 to 6 flowers.
Flower: The sepal is longer but narrower than the petal. The sepal is pale green. The petal is also pale green but has a central purple dotted line. The side-lobe is low, rounded and has purple markings. The mid-lobe is broad and its end curves downwards. The are two longitudinal purple markings near the tip. On its basal part are two short fleshy keels.
The column is green with some purple markings.

Cymbidium lancifolium is widely distributed, extending from India to Japan and southwards to Indonesia and the surrounding islands. In Peninsular Malaysia, the species inhabits the montane forest areas.

Cymbidium lancifolium, a montane species with pale green flowers.

The flowers of *Cymbidium simulans* are similar to *Cymbidium finlaysonianum* (see page 45).

Cymbidium simulans Rolfe.

Plant: is generally smaller than *C. finlaysonianum* though the growth habit is similar.
Leaf: is narrower and shorter but appears similar to that of *C. finlaysonianum*.
Inflorescence: is pendulous with many flowers. Compared to *C. finlaysonianum* it is smaller, shorter and more slender.
Flower: The sepal is slightly longer than the petal, but they are otherwise about the same size. The apices of both are pointed. There is a median deep purple line while the edge is yellowish white. This colour combination makes the flower attractive. The side-lobe of the lip is longer than the column. Its apex is acute and points forward. On the side-lobe are many purple stripes. The mid-lobe is tongue-shaped and its terminal half curls downwards. The apex is deep purple, but inwards are purple stripes. At the base between the side-lobes, are two yellow keels and these form a continuous furrow all the way to the column-foot. The furrow is constricted midway.
The column is about three-fourths the length of the dorsal sepal. It is solid purple. The anther cap is yellow.

Cultivation

Some Cymbidium species are commonly known to orchid growers. They are being grown in the home garden as epiphytic plants. They are decorative and are interesting collectors' items. Cybidiums are very hardy plants. For growing the epiphytic type the clump should be tied onto a strong wooden support. It should be placed at a sufficient height for the long pendulous inflorescence. Once established the plant needs minimum attention.

For the terrestrial type a pot filled with loose soil should be used. This type of Cymbidium needs less space and is not as bulky as the epiphytic types. Furthermore, the flower is erect and seems more attractive.

The furrowed lip of *Cymbidium simulans*.

4mm

FLICKINGERIA

The members of this genus were once placed in the section Desmotrichum within genus Dendrobium while in some other books this genus was referred to as Ephemerantha. In the author's correspondence with Seidenfaden, the latter has pointed out that Ephemerantha is not a valid name and the genus should rightly be called Flickingeria.

These nomenclatural problems started when Blume in 1825 established the genus Desmotrichum, comprising of 12 species from Java. Lindley in 1830 transferred all of Blume's Desmotrichum species to the genus Dendrobium. Using the same guideline in Holttum's book these were referred to as Dendrobium. However, one must agree that *D. plicatile*, for example, cannot be a Dendrobium! Two characteristics help distinguish it from Dendrobium. First, it has a wiry rhizome which supports an enlarged pseudobulb and which may be branched. The plant is a scrambler and grows in clusters. Secondly, the edge of the lip of the flower is much folded. For these reasons there is no doubt that this group of orchids deserved a new generic name.

In January 1961 Hawkes proposed the new name, Flickingeria, for the genus, noting that the original name, Desmotrichum, had been used elsewhere since 1845. About the same time (June 1961) Hunt & Summerhayes proposed another new name for the genus. This they called Ephemerantha. They noted that Hawke's suggested name did not comply with the international code. Unfortunately, this opinion is not correct. Nevertheless, this gives rise to some confusion. The genus Flikingeria consists of some 60–70 species distributed throughout mainland Asia and the islands of Indonesia, Philippines, New Guinea and the Pacific.

Flickingeria fimbriata (Bl.) Hawkes

Previous names:
Desmotrichum fimbriatum Bl.
Desmotrichum plicatile Lindl.
Dendrobium plicatile Lindl.

Rhizome: is almost round, pencil-sized, short and branching. At the end of the rhizome or stem is the flattened pseudobulb which is slightly grooved. There is a tendency for the rhizome to produce roots at the node.
Leaf: is smooth and leathery. It tapers at both ends.
Flower: is solitary borne at the end of the pseudobulb. It is orientated upside down when in full bloom.
The sepal is slightly broader and larger than the sepal. Both are pale yellow with purple streaks or dots on the inside. Outside they are not spotted. The mentum or chin is long, about three-fourths the length of the flower stalk and is angled about 80° in relation to the stalk. The mid-lobe is prominent. It is enlarged with folded edges. There are 3 wavy ridges running all the way down the throat. The mid-lobe closes like an umbrella to form a pointed head. This happens when the flower dries up. The side lobe is triangular and points forward. Inside are purple streaks.
The column is short and straight. It appears to be in line or a "continuation" of the flower stalk.

F. fimbriata is native to Peninsular Malaysia and the surrounding regions: Java, Celebes and East Malaysia. In Peninsular Malaysia it is found in many localities in the lowlands and at medium elevations on the hills.

Cultivation

The plant is not found in cultivation. It can be grown as a scrambler on rocks on the ground under the shade. The plant appears to be hardy.

The flower of *Flickingeria fimbriata* is solitary and oriented 'upside down'.

GRAMMATOPHYLLUM

This is a genus of about six species which inhabits the lowland forests of Indochina, Thailand, Malaysia, Indonesia, New Guinea and the Philippine Islands. Two species, *G. speciosum* and *G. peliiflorum* are native to Peninsular Malaysia.

In the earlier days there was some confusion as to the naming of plants in this genus. Linnaeus called it Epidendrum while Spreng, Swartz, and Wight referred to it as Vanda, Cymbidium and Pattonia, respectively. It was not until 1825 that Blume officially described it as Grammatophyllum using *G. speciosum* as a type species. This generic name is derived from two Greek words, *gramma* = mark, character or letter; and *phyllum* = a leaf.

The abnormal flower of the tiger orchid which appears at the lower part of the flower spike.

Grammatophyllum speciosum Bl.

Previous names:
Grammatophyllum fatuosum Lindl.
G. macranthum Reichb.f.
G. wallisii Reichb.f.
G. giganteum Bl.
G. sanderianum Hort.
Pattonia macrantha Wight

Plant: is the largest species in the genus. The pseudobulb grows to the size and height of the sugar-cane plant. Holttum noted two vegetative forms, though both produce identical flowers:
(a) Plant with long, fleshy pseudobulb and many leaves
(b) Plant with rather short but proportionately thicker pseudobulb and few leaves at the apex.
Leaf: is long, very narrow and leathery.
Inflorescence: appears out of the clump at the base of the stem like the bamboo shoot. The inflorescence can be up to 2 metres in length and has many flowers. The flowers on the lower part of the inflo-

Grammatophyllum speciosum, the tiger orchid.

rescence are usually sterile having some parts missing. For example, it may have only the sepals and petals without the lip and column. The lower flowers are arranged far apart but towards the apex the flowers are well-placed forming a nice cluster. The flower stalk is long and thick.

Flower: Each flower is similar to the Cymbidium except that in this case there are 3 low keels on the lip. They also differ in the siting of their pollinia.

The sepal and petal are about the same size. Both are spreading and are greenish yellow blotched with intense dark-brown spots. The lip is about half the length of the sepal. It has red-brown markings and is hairy within. The mid-lobe has three low keels. The side-lobe is erect, and has many parallel brown stripes on the inner side.

The column is pale green, white underneath with purple spots. An outgrowth on either side of the base of the column joins with the base of the lip to form a cup-shaped hollow.

In nature *G. speciosum* is an epiphyte. It is found growing stems hanging down from trunks of trees or palms along streams or rivers of the lowlands. Aerial roots can be seen clasping firmly onto the host-plant, in their effort to support the enormous weight of the entire plant. The species can be found in Thailand down through Malaysia, Indonesia and the Philippines. In New Guinea there is *G. papuanum* which according to some is not a distinct species but a variation of *G. speciosum*.

Cultivation

The plant can be treated as a terrestrial orchid. Because of its enormous size, it is advisable to plant it in the ground, rather than in a container, on a raised bed made from broken bricks, stone, etc. This is to ensure good drainage. The root should be covered with compost or friable soil. Animal manure should be added once in a while. The plant needs plenty of sunlight and does well with good watering.

G. speciosum is known locally as the Tiger Orchid. In the Philippines it is referred to by such names as: Giant Orchid, Queen Orchid or Sugar-cane Orchid. The specific name of this plant is derived from the word *speciosum*; meaning handsome or showy. Indeed the name is most appropriate. A plant in bloom is most impressive and showy! Bryan reported in the American Orchid Society Bulletin (1947) that a plant cultivated in his garden in Hawaii grew into a clump of about 9 feet in diameter and the plant was 10–12 feet high after 16 years. It produced well over 10,000 flowers at first bloom. Such show of beauty, however, does not come too frequently to any grower, for *G. speciosum* does not flower freely. The plant is not expected to give its first bloom until after 7 to 9 years of cultivation.

GEODORUM

This is a genus of about 10 species distributed from India to Japan, Australia and the neighbouring islands. Two species are native to Peninsular Malaysia: *G. densiflorum* and *G. citrinum*. Both these species can be easily mistaken for each other. The distinctive feature of the genus is its curved inflorescence giving the shape of a "shepherd's rod".

The name of the genus is derived from Greek, *geo* meaning earth, and *doron* meaning gift. This "gift of the earth" probably refers to the terrestrial habit of the plant.

Geodorum densiflorum (Lam.) Schltr.

Previous name:
G. purpureum R.Br.

Pseudobulb: is almost round and is placed in the ground.
Leaf: is broad and thin with 5 to 7 veins.
Inflorescence: is thick, fleshy and curved at the end like a shepherd's rod. At the end of this inflorescence, the flowers are borne clustered close together and facing the ground. They open in succession.
Flower: Each flower is subtended by a long bract. The petal is whitish yellow and is broader than the sepal. The petal is broad in the middle and tapers to a blunt apex. The sepal and the petal do not spread out, thus each flower appears about three-fourths open only. From side-view, the lip has the shape of a long bathtub. It is attached about perpendicular to the column-foot and is broad and concave. Its terminal end is brown with a patch of yellow. The apex of the lip is broad and straight and not notched.
The column is short and is about half the length of the lip.

G. densiflorum is found in the regions of Burma down through Malaysia and Indonesia. In Malaysia it inhabits the open grassland

The bathtub-shaped lip of *Geodorum densiflorum*.

6.5mm

The curved inflorescence of *Geodorum densiflorum* is shaped like a shepherd's rod.

Flowers of *Geodorum densiflorum* viewed from the underside.

and waste ground.

In papers written by Seidenfaden, the name *G. purpureum* has been changed to *G. densiflorum*. According to him this genus "Geodorum is indeed the most complicated one in South East Asia", and he has "written some 25 to 30 pages about this genus without getting near to any reasonable result." For this reason this author shall not go any further than accept this suggested name.

Geodorum citrinum Jacks

This is a species which is very much similar to *G.densiflorum* in most aspects. Therefore it is not surprising if the species are often mistaken for each other. However, careful comparison of the flowers of the two species will bring out the following two characteristics which differentiate them from each other.

1) The petal of *G. citrinum* is broader and more rounded. It is not broad midway and tapering at both ends as in *G. densiflorum*.
2) The apex of the lip tapers to a narrow end and is slightly notched.

G. citrinum is distributed in Burma and Thailand and extends south up to the northern states of Peninsular Malaysia.

Cultivation

Geodorum is not cultivated except as a botanical specimen. It should be grown in well-drained soil in a shady place or even among grasses.

HABENARIA

Habenaria is one of the largest genera of terrestrial orchids. It is found particularly in Brazil, tropical Africa, the north temperate zone and the South East Asian regions. In his book, Holttum listed some 14 species of Habenaria that are native to Peninsular Malaysia. However, in the light of nomenclatural changes eight of these species listed ought to rightly belong to other genera: Platanthera, Pecteilis, and Peristylus. Among the remaining Habenaria species, *H. rhodocheila* and *H. carnea* are unusually beautiful and distinctive. They are different in character from most other cultivated orchids. The other attractive species which has not been described in Holttum's book is *H. medioflexa*.

The generic name, Habenaria, is derived from the Greek word *habena* = reins, possibly referring to the two long, protruding anther canals.

The feather-like lip of *Habenaria medioflexa*.

5mm

Habenaria medioflexa Turill

Plant: is about 45 cm high. It produces tubers measuring about 3 cm x 1 cm.
Leaf: is stalkless and its base clasps the stem. There are six to eight leaves per plant. The largest leaf measures about 20 cm long by 5 cm wide.
Inflorescence: is terminal. It carries some eight to 14 flowers. Each flower is subtended by a green pointed bract.
Flower: The dorsal sepal and petal are joined to form a hood. The lateral sepal is broadly crescent-shaped with two distinct veins. The mid-lobe is white, finely divided like feathers. The spur is long and is boomerang-shaped.
The plant flowers sometime between September and October, as well as March to April.

H. medioflexa is said to be endemic to Thailand where it can be found in elevated areas of the east, north and south.

The plant illustrated in this book, however, was found in the

Habenaria medioflexa, a species thought to be only endemic to Thailand.

foothills of Kedah, a northern state of Peninsular Malaysia. The collection was made by Mr. Cheang Kok Choy, ex-Head of the Penang Botanical Gardens.

The author sent a preserved specimen to Seidenfaden for identification. According to Seidenfaden there is little doubt that this species is *Habenaria medioflexa* Turill. The plant was originally described from a specimen collected in Peninsular Thailand. *H. medioflexa* was found several times in Thailand by various people, and Rolfe, not knowing of Turill's publication, described it again as a new species which he called *H. trichochila*.

Habenaria rhodocheila Hance

Previous names:
H. xanthocheila Ridl.
H. pusilla Rchb.f.

Plant: can be 30 to 40 cm high.
Leaf: is green but has a fine network of darker veins. There are 5 to 6 leaves per stem. With height, the size of the leaf gets smaller, grading into narrow bract-like structures.
Inflorescence: can be 4 to 6 cm long, and has numerous flowers. The flower has a long stalk with 5 to 6 parallel ridges. It is subtended by a bract.
Flower: The dorsal sepal forms a conspicuous green hood over the column. The hood is 5-veined and is triangular if seen from the front. The two lateral petals are thin and narrow and are horizontally placed in line with the lower side of the hood.

The two lateral sepals are twisted. The lip is bright orange yellow. It is broad and deeply 3-lobed. The spur is very long and extends more than the length of the flower stalk. The column is short, about half the length of the hood. It is connected to a protruding arm on each side of the column. In this structure is placed the anther with its long filaments. The filament terminates in a sticky, glistening red papilla. This papilla sticks to an insect when it visits the flower in trying to get the nectar from the spur.

There are two stigmas, each in the form of an elongated club-shaped process placed on either side of the base of the column. At the base of the stigma is a small round gland.

H. rhodocheila is found in regions extending from South China, Indo-China downwards through Thailand and the northern region of Peninsular Malaysia. It inhabits open or partially exposed, moist and rocky places of elevated forest.

Cultivation

Habenaria grows well on loose medium, particularly leaf mound and dead wood. It loves the shade and does well under trees kept away from direct sunlight. After flowering, the leaves gradually dry out and new plantlets arise from the tubers at the base of the plant.

For some Habenaria species the tubers may need a certain period of rest before new sprouts are formed. In such species this resting period is important so as to enable strong new plantlets to sprout. During the resting period the tubers should be kept rather dry and watered sparingly.

In Malaysia, and possibly also true elsewhere, Habenaria is rarely planted and is, often unknown to most orchid enthusiasts. Spathoglottis and Arundina are more preferred terrestrial plants to cultivate than Habenaria. The lack of popularity for Habenaria could be due to many reasons. This genus itself is not often known to the collectors and beautiful species are not easy to obtain. It is deciduous and the tuber may be dormant for a while, and such special cultural requirements are often unknown to most growers. The flowers, though distinctive and unique, are not large and long lasting as the Cattleyas, Dendrobiums or Vandas. In spite of all these reasons, Habenaria could be an interesting item for serious collectors to possess.

The flowers of *Habenaria rhodocheila* from Malaysia are bright orange in colour.

KINGIDIUM

Kingidium is represented by one species in Peninsular Malaysia: *Kingidium deliciosum*. The plant is better know by its former name: *Phalaenopsis decumbens*. Indeed the genus Kingidium has been in a limbo for quite some time.

The first illustration of this species was made by Griffith, and he labelled it as *Aerides decumbens* Griff. Since then the species ran into nomenclatural problems and it has been described by different people by different names. It was once called a Doritis. J.J. Smith included it in the genus Phalaenopsis. In Malaysia, as in Holttum's book, this species was called *Phal. decumbens* (Griff.) Holtt. Rolfe in 1971, however, separated it from the Phalaenopsis and created a new genus called Kingiella. This was done in memory of Sir George King in recognition of his work on the Indian orchids. More recently, the name of the genus underwent a slight "plastic surgery" and the species was called *Kingidium decumbens* (Griff.) P.F. Hunt. One would think that that was a happy ending and the species rests in peace in that slot. However, Sweet in 1969 pointed out that the original illustration of *Aerides decumbens* by Griffith failed to meet the requirements of the rules of the International Code of naming plants, and therefore the illustration, was not acceptable as a recognised work. It followed therefore that its name *decumbens* had to be rejected. So by the rules of modern man, the species was again given a new name *Kingidium deliciosum* (Rchb.f.) Sweet.

Kingidium deliciosum (Rchb.f.) Sweet

Previous names:
Phalaenopsis decumbens (Griff.) Holtt.
Kingiella decumbens (Griff.) Rolfe
Doritis winghtii (Rchb.f.) Benth.

Plant: is small.
Leaf: is narrow and long. It is rather thin.

Inflorescence: is unbranched and carries a few flowers.
Flower: is small. The sepal is broader than the petal. The dorsal sepal is broader than the two lateral sepals. All are white in colour except for the lower portions of the lateral sepals which have some purple dots. The lip is pinkish. The side-lobe has stripes of pink with two appendages projecting inwards to meet at the middle. The mid-lobe has a forked tongue-like structure at the base.
The column is short. It is light pink in colour.

Kingidium deliciosum is widely distributed in the area ranging from South India, Sri Lanka, Burma, Thailand, Indochina, Malaysia to Indonesia and the Philippines. In Malaysia, the plant can be found growing on shady tree trunks in the lowlands. The mangosteen (*Garcinia mangostana*) trunk seems to be most attractive to the plant.

Cultivation

The plant can be grown like Phalaenopsis tied to a tree trunk, tree-fern or coconut husk. Grow it in the shade under humid but well aerated conditions.

The phalaenopsis-like flower of *Kingidium deliciosum* showing the forked mid-lobe.

MALLEOLA

This is a genus of about 30 species distributed throughout the region ranging from India to New Guinea. Vegetatively, the plant is similar to Cleisostoma. Though the flowers appear alike, they differ in their fine structures; especially the lip and column. Malleola has two pollinia while there are four in Cleisostoma.

The generic name Malleola is derived from the Latin *mallaelus* meaning little hammer. It is also named possibly because the flower has a prominent hammer like spur. The lobes of the lip are very small and insignificant. It was suggested that because of this, the moth visiting the flower is unable to land on the lip. Instead, the broad spur which bends backwards acts as a landing stage for the visiting insect.

Holttum listed 7 species of Malleola which are native to Peninsular Malaysia, and he suggested that there may be many more which remain unrecorded.

Malleola insectifera (J.J.S.) J.J.S. et Schltr.

Previous name:
Saccolabium insectiferum J.J.S.

Plant: is small, resembling the Cleisostoma.
Leaf: is broad, leathery and has a blunt, unequal apex. The edge is slightly wavy.
Inflorescence: is long, slender and hangs down below the leaves. The flower is sparsely arranged. Each flower is small and is subtended by a small, brown bract.
Flower: The dorsal sepal is about semi-circular, has a sharp apex and forms a hood over the column. It is yellowish brown with a yellow median band. The lateral sepal and petal are pushed vertically backwards. They are yellowish brown with a yellow median band. The lip is characteristic of the genus. The side-lobe is erect, triangular, with one side pointing forward and slightly in-curved. The mid-lobe is narrow, pointed, and coils over to form a semi-

The duck-like flower of *Malleola insectifera*.

2.5mm

circle. It is yellow at the tip and dark at the base. The spur is broad and bends backwards to meet the flower stalk. It is slightly flattened laterally. The upper tip of the spur is purple while the rest is pale yellow. The spur is the most prominent feature of the flower. At close view, it resembles a duck swimming in the water. The column is short, fleshy and has two horns in front on which are inserted the two pollinia. The upper half of the column is white while its lower half is purple.

Malleola sp.

Plant: is small and appears very similar to *Cleisostoma subulatum*.
Leaf: is leathery and appears flat.
Inflorescence: is long, unbranched with numerous small flowers.
Flower: The dorsal sepal is pale yellow with some purple lines. It forms a hood over the column. The lateral sepals are yellow with two broad brown lines. About one-fourth of the length from the apex it curves inwards. The petal is yellow with blotches of purple and is narrower than the sepal. The sepal and petal are spreading. The lip is spurred and this is most prominent. The side-lobe appears non-existent. The mid-lobe takes the form of a sharp projection. At the entrance of the spur are two purple markings. The spur is long, narrow and gently curved. The apex is blunt.

Cultivation

Because of its relatively small and insignificant flowers, the plant is not grown by orchidists. It is, however, an interesting specimen and is well worth cultivating. The flower is rather attractive and its structure unusual. The plant grows well when tied to a tree trunk in the shade.

PAPHIOPEDILUM

This genus of Asiatic Lady's Slipper has about 50 species which are distributed from the regions of the Himalayas through to the South East Asian regions. Most of them are inhabitants of the cool montane areas. There are 7 species native to Peninsular Malaysia. Of these only 2 species, *P. barbatum* and *P. niveum* are often heard of; the rest seem rare to collectors.

Paphiopedilum species are in great demand in the temperate countries. As such they have been stripped from their native habitats and exported.

The generic name of Paphiopedilum is derived from the Greek *paphia,* i.e. of Paphos, epithet of Venus, and *pedilon* = sandal. The name Sandal of Venus is an obvious reference to the slipper-shaped lip.

A select plant of the Malaysian *Paphiopedilum niveum*.

Paphiopedilum niveum (Rchb.f.) Pfitz.

Previous name:
Cypripedium niveum Rchb.f.

Plant: is a terrestrial with a short stem.
Leaf: is oblong, greyish dark green with pale mottling. The underside of the leaf is brownish purple.
Inflorescence: arises from the heart of the shoot. The spike is long, bearing one or two almost white flowers.
Flower: The dorsal sepal is almost round and is pure white. The two lateral sepals unite to form the synsepalum and this is about $2/3$ the length of the dorsal sepal. The petal is narrower than the sepal. It is bright white with tiny dots on the basal half of the petal. The lip forms a pouch with a rounded base. In some plants there are tiny dots throughout the lip.

The staminode (modification of the column, in other genera) is broad with a big yellow patch.

Paphiopedilum niveum inhabits the low limestone areas stretching from Pulau Langkawi to Perlis, northern Kedah and Kelantan. It grows in decayed vegetation found among the rock crevices of limestone outcrops. In Pulau Langkawi, the plant even grows at sea-level. The distribution of *P. niveum* extends northwards into Peninsular Thailand.

Paphiopedilum barbatum (Lindl.) Pfitz.

Previous name:
Cypripedium barbatum Lindl.

Plant: is a terrestrial bearing two rows of leaves. The stem is short and the leaves arranged closely together.
Leaf: is leathery, strongly tessellated with light and dark bluish green colour.
Inflorescence: is long, covered with fine, short hairs. It bears one, rarely two flowers.
Flower: the dorsal sepal is large and conspicuous. It has distinct lines of green and purple against a white background. The petal spreads horizontally (but in other variants, it forms an inverted V). The petal also has green and purple lines. Along the margin are fine hairs. The upper margin of the petal has 3 to 4 black warts which are randomly arranged. The lip forms a broad pouch and is purple.

Paphiopedilum barbatum is found in several localities in Peninsular Malaysia. Popularly known areas are: Penang Hill, Gunung Jerai and Gunung Ledang (Mt. Ophir) where the species was first noted by Hugh Cuming in 1838. Alphonso (former Curator of Singapore Botanic Gardens) reported the presence of this species on Gunung Blumut, Gunung Ayam and Gunung Tebu in Trengganu and Bukit Yong in Kelantan. The species can also be found in Thailand.

The flowers of *Paphiopedilum barbatum* show distinct variations in specimens collected from different localities. The accompanying photographs illustrate these variations.

A *Paphiopedilum barbatum* plant that flowers in lowland conditions.

Cultivation

Except for *P. niveum*, the other species of Pahiopedilum are montane species. This implies that the plant would not grow well under lowland conditions. Though this ought be true, *P. barbatum* collected from certain localities could do well in the lowlands. It needs cool, shady conditions. The plant can be grown in a well-rotted leaf litter or a thick mat of coconut fibres placed in a wooden basket. For *P. niveum*, some lime must be added to the growing medium.

The species of Paphiopedilum are one main target of "robbery" by the commercial orchid exporters. There is a definite need to conserve them!

Examples of the variations found in *Pahiopedilum barbatum* collected from various localities in Peninsular Malaysia. Note the differences in the purple and green colouration.

PHALAENOPSIS

The genus Phalaenopsis comprises about 70 species. They are distributed from the regions of the Himalayas through to Thailand, Indo-China, Malaysia, Indonesia, New Guinea and Australia. Northwards, its distribution extends upwards to Taiwan and Southern China. The Philippines appears to be the primary centre of the genus, having recorded 42 species and some 36 varieties.

Phalaenopsis is known for its graceful, arching spray of beautiful flowers. On the plant, the flowers appear like butterflies with wings outstretched. In the Philippines this flower is known as *mariposa* meaning butterfly. In Indonesia, the Phalaenopsis is known as *anggerik bulan* or the moon orchid, possibly comparing it to the romantic full moon. It was in Java in 1885 that *anggerik bulan* was named Phalaenopsis. Blume gave it this name after having studied a specimen of *P. amabilis*. The name Phalaenopsis is derived from two Greek words, *phalaina*, = moth; and *opsis* = appearance.

The Malaysian Phalaenopsis, however, does not resemble a butterfly or have that graceful spray. Those that are native to this peninsula have short or flattened inflorescence with one or two flowers opening at a time. The flower is also small and star-shaped.

Five species have been recorded: *P. violacea, P. cornu-cervi, P. fuscata, P. sumatrana* and *P. appendiculata*.

Phalaenopsis cornu-cervi (Breda) Bl. and Rchb.f.

Previous name:
Polychilus cornu-cervi Breda

Stem: is short.
Leaf: is narrow and long. It is pale green. There are few leaves on each plant.
Inflorescence: is branched and flattened. It is long and usually hangs down.
Flowers: open in succession, one or two at a time. The plant stays in bloom for months. The sepal is narrow and tapering. It is light

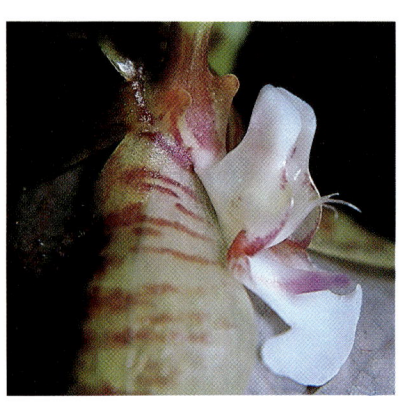

The appendages on the lip of *Phalaenopsis cornu-cervi*.

11mm

yellow with brown bars or spots. Colour intensity varies with different specimens. The petal is shorter and narrower than the sepal but is about the same colour as the sepal. The lip has a complicated structure. The two side-lobes are erect and have brown stripes on the inner side. The mid-lobe has a hollow depression. From its upper rim arises a yellow tongue structure and from the lower rim a broad three-pronged tongue. The terminal portion of the mid-lobe is white and is broad like an arrow-head. Midway along this lobe arises another mauve, tongue-like structure.
The column is short and is winged towards the apex. At the base of the column on each side are two ridges.

P. cornu-cervi is found in Burma, Thailand, Peninsular Malaysia, Borneo and Sumatra. It grows in areas which are more exposed than where P. violacea is found.

The side-lobes and tongue structure of Phalaenopsis cornu-cervi.

5mm

Rare dark chocolate flowers of a Phalaenopsis cornu-cervi.

Phalaenopsis fuscata Rchb.f.

Previous names:
Phalaenopsis kunstleri Hk.f.
Phalaenopsis viridis J.J.S.

Stem: is short
Leaf: is about three times as long as it is broad. It is rather thin.
Inflorescence: is long, usually with several lateral branches.
Flowers: usually open all at once. There are three or four towards the end of each branch. The sepal is broader and longer than the petal. The apex is rounded. It is yellow flushed with brown towards the basal end. The lip is shorter than the sepal and is also complicated. The side-lobe is erect and is yellow with purple bars on the outside and purple dots on the inside. At the base between the side-lobes are three structures: two distinct calli of white with purple dots, in front of which is a hook-like structure on each side, and towards the base of the mid-lobe is a pointed two-pronged tongue-like structure. The mid-lobe is shaped like a rounded Chinese hand fan. Running through vertically in the middle is a keel. The mid-lobe is yellow in colour with 4 broad brown lines parallel to the keel.
The column is short and fleshy.

P. fuscata is found in Peninsular Malaysia, Borneo and Sumatra. It grows in shady areas on tree trunks, often by streams or rivers.

The rare all-yellow form of *Phalaenopsis fuscata*.

Phalaenopsis violacea Teijsm and Binn.

Stem: is short.
Leaf: is broad, succulent and shining dark green.
Inflorescence: is flattened, zig-zag with alternate bracts.
Flowers: open in succession, one or two at a time. A spike can have 6 to 10 flower buds. The sepal is slightly broader than the petal. Both are mauve, entirely or partly. The colour intensity and pattern, however, vary with plant specimens. The apices of the sepal and petal are pointed and tinted green.

The complicated lip is not hinged to the foot-column. The two side lobes are erect with edges curved inwards to meet one another. These lobes are coloured mauve, yellow and white. There are some prominent dark mauve spots on the inner side or the erect lobes. At the base of the lobes is a callus. The mid-lobe is tongue-like and is keeled. At its inner end is a two-pronged tongue structure. Towards the apex on the underside is a notch giving it the appearance of a chin.
The column is about three-fourths the length of the sepal.

P. violacea is found in Peninsular Malaysia, Borneo (the flower of the Borneo type is bigger and rounder. It is more valued by collectors.) and Sumatra. It grows at low altitudes in shady areas, on tree trunks, overhanging streams or rivers.

The tongue-like appendages found on the lip of *Phalaenopsis violacea*.

5mm

The two erect side-lobes of the lip.

2.5mm

The Peninsular Malaysian *Phalaenopsis violacea*.

Cultivation

Phalaenopsis grows well if attached to pieces of wood or tree trunks, coconut husks or tree ferns. It needs full shade. It should not be grown exposed to the full sun. The leaves will get burnt. Since the plant has no pseudobulb and is succulent, growing it in an area with high humidity but good aeration, is ideal.

The all-white or alba form of *Phalaenopsis violacea*.
The plant is highly prized and much sought after.

PODOCHILUS

Podochilus is a genus of about 75 species of mostly small, often moss-like epiphytes. These orchids are generally unknown in cultivation. They can grow into a mat of green fine-looking delightful plants found on moist tree trunks, rocks and boulders. The flower is tiny and botanically interesting. Podochilus has a wide range of distribution extending from Sri Lanka to New Guinea. Four species of Podochilus are native to Peninsular Malaysia.

Podochilus tenuis (Bl.) Lindl.

Previous names:
Apista tenuis Bl.
Podochilus acicularis Hk.f.

Stem: is long and slender.
Leaf: is sharp and narrow, held at a very acute angle to the needle-like stem.
Inflorescence: is terminal with 2 to 3 flowers.
Flower: The dorsal sepal is white with faint purple spots. At the apex of each of the lateral sepals is a prominent purple spot. The lateral sepals and the lip meet to form a swollen, rounded mentum like the scroto-sac. The sepal is narrower and shorter than the petal. It is white without any spot. The lip has two purple dots at the apex. The blade of the lip is bluntly triangular. At the inner end are two lateral appendages. The lip is connected to the column by a rounded structure.
The column is short and broad. The anther is on the upper side of the column and below is the stigma which appears framed within a rectangular box.

P. tenuis is found on wet mossy trunks and rocks. It occurs throughout Peninsular Malaysia in the lowlands up to an elevation of 2,000 feet. The species is also native to Indonesia and East Malaysia.

Cultivation

Podochilus is not cultivated and is unknown to most orchid growers. It grows well when tied to wood pieces and placed in moist, shady areas.

The erect, branched inflorescence of *Pomatocalpa kunstleri*.

POMATOCALPA

In the earlier days, some members of this genus were placed in various genera such as: Cleisostoma, Saccolabium, Sarcanthus etc. instead of Pomatocalpa.

The generic name, Pomatocalpa, is derived from the Greek words *poma* = a cover or lip, and *kalpa* = a jar or pitcher. This is obviously in reference to the structure of the labellum, although it appears that a "flower with a lid covering a jar-like lip" is not quite correct to describe the genus. Generally, the flower of Pomatocalpa is small, rather succulent, with spreading sepals and petals. The spur takes the shape of a pouch or pitcher. At the back wall of the spur is a tongue and in some species it protrudes out of the opening.

Pomatocalpa is a genus of some 30 species distributed throughout the Asia-Pacific region covering areas from Sri Lanka to Samoa.

Six species are native to Peninsular Malaysia *P. latifolium, P. kunstleri, P. spicatum, P. setulense, P. arachnanthe*, and *P. elongatum*.

Pomatocalpa kunstleri (Hk.f.) J.J.S.

Previous names:
Cleisostoma kunstleri Hk.f.
Saccolabium pubescans Ridl.

Stem: is rather short.
Leaf: is narrow an long, dull grey-green. Its apex is broadly rounded and unequally bilobed.
Inflorescence: is much branched and erect. The end of the inflorescence droops down. Each spike carries numerous small pinkish-white flowers.
Flower: The sepal is narrow with rounded apex. It is broader than the petal. Both are pinkish with some darker stripes. The mid-lobe of the lip is short and triangular. It bends over vertically. The side-lobe is erect and fleshy. The spur is slim and is about as long as the dorsal sepal. The opening to the spur is narrow. The bottom of the

spur is rounded. Within the spur arises a two pronged tongue about a quarter length down the back wall of the spur. The front wall of the spur is thick but not very fleshy.
The column is short.

P. kunstleri can be found in the islands of Indonesia, and both east and west of Peninsular Malaysia. It is commonly found in shady places of the lowlands.

Pomatocalpa latifolium (Lindl.) J.J.S.

Clusters of flowers appear at the ends of the branched inflorescences of *Pomatocalpa latifolium*.

Previous names:
Cleisostoma latifolium Lindl.
Saccolabium hortense Ridl.

Stem: can be short or long
Leaf: is narrow and long. It is slightly shining but often yellowish green . It has a leathery texture. The apex of the leaf is rounded and is bilobed.
Inflorescence: is branched, carrying clusters of flowers at the end of each branch. Each flower is small.
Flower: the sepal is slightly longer and broader than the petal. Both are yellow in colour with a broad lining of reddish-brown along the edge. The apex of the sepal or petal is rounded. The lip is yellow. The mid-lobe is broadly triangular and it bends over, its tip touching the spur. The side-lobe is short and triangular in shape. The spur is prominent. It appears like a round-bottomed pitcher. It is yellow with reddish-brown spots. The front-wall of the spur is thick and fleshy. An erect tongue reaches out to the mouth of the spur.
The column is short.

P. latifolium is distributed in areas of Indonesia, Malaysia and the Philippines. It is a lowland plant that grows naturally in exposed areas as an epiphyte on trees.

Pomatocalpa setulense (Ridl.) Holtt.

Previous name:
Saccolabium setulense Ridl.

Stem: is short with leaves compactly arranged.
Leaf: has a slightly wavy margin.The apex is unequally bilobed and rounded.
Inflorescence: with 3 to 4 branches often appears below the leaves. There are many tiny flowers well arranged in clusters.
Flower: Sepal is broader and longer than petal. Both are clear greenish yellow in colour. The sepal and petal curve forward not spreading, giving the appearance that the flower has not fully bloomed. The lateral sepals curve around the rounded lip. At the base of each lateral sepal is a big reddish spot. The lip which forms the spur is

yellow but at the rim it is faint red. The side-lobe is non-existent. The mid-lobe is rounded and curves backwards. A section through the spur shows the presence of a membranous tongue arising quarter way down its back wall. In front, the tongue touches the thick callus of the front wall.

The column is short and broad.

*P. setulense i*s found on trees and rocks, especially in limestone areas of north Peninsular Malaysia and southern Thailand.

In *Pomatocalpa setulense* the membranous tongue arises from the back wall of the spur. The front wall forms a thick callus.

1mm

Pomatocalpa setulense with big reddish spots on the lateral sepals.

2mm

Pomatocalpa spicatum Breda.

Previous names:
Cleisostoma uteriferum Hk.f.
Saccolabium uteriferum Ridl.
Saccolabium hobsonii Ridl.

Stem: is very short.
Leaf: is long and leathery. It has many edges.
Inflorescence: can be single or branched. The inflorescence is short with numerous closely arranged flowers. It appears below the leaves, pointing downwards.
Flower: Each flower is small. The sepal and petal are short and broad. They are all yellow in colour with some pink at the base. The lateral sepals may also have some pink markings. These sepals are concave and bend forward to hug the lip. The mid-lobe of the lip is broad, rounded and tapers to a blunt apex. The spur is as short as the flower stalk. It is broad with a rounded bottom. The opening of the spur is as wide as the width of the spur. A short broad tongue arises about midway down the backwall of the spur.

P. spicatum is found as an epiphyte on trees growing in shady places throughout Malaysia and the Indonesian islands.

Cultivation

Pomatocalpa is generally not cultivated by hobbyists. Most of the species have inconspicuous flowers. Possibly, the only species that is attractive is *P. latifolium*. It can be grown tied to a wooden support and placed in a shady place. The plant can also stand in a more exposed position.

Downward growing inflorescence of *Pomatocalpa spicatum*.

PORPHYRODESME

The genus Porphyrodesme was established by Schlechter in 1913. The name was first used to describe *Porhyrodesme papuana* Schlter, a species inhabiting the high mountain forest of New Guinea. Since then, the genus seemed to have been forgotten and Porphyrodesme is regarded as a rare one-species genus.

Garay, after having carefully studied the Renantheras, suggested that some members in this genus should correctly belong to the Porphyrodesme. Hence, what is commonly known as *Renanthera elongata* (possibly also *R. moluccana*) should now be known as *Porphyrodesme elongata*.

The generic name, Porphyrodesme originated from two Greek words: *porphyra* meaning purple, and *desme* meaning bundle. Together the words possibly refer to the reddish inflorescence with clusters of scarlet flowers.

Porphyrodesme elongata (Bl.) Garay

Previous names:
Renanthera elongata (Bl.) Lindl.
Renanthera micrantha Bl.
Aerides elongata Bl.
Saccolabium reflexum Lindl.

Stem: is long and climbing.
Leaf: is bilobed and has a purplish sheath.
Inflorescence: is long with several alternate branches. It has numerous, small flowers.
Flower: The stalk is as long as the sepal. The entire flower is yellowish or orange red with red spots throughout. The sepal is longer and broader than the petal. The basal half of the sepal is broad and flat while the terminal half rolls back. The lateral sepal is narrow at the base and widens in the middle. The apex is round. The midportion of each lateral sepal has a protruding flange on its inner edge. These two structures point towards each other. The lip is at-

tached to the column-foot. The side-lobe is erect and rectangular. The end of the mid-lobe curves under. The spur is short with a rounded apex and is slightly compressed on the sides. This effectively closes the opening of the spur.

The column is broad and short. It is entirely red towards the terminal half while the basal end is white.

Porphyrodesme elongata is also native to Borneo and the Indonesian Islands. In Peninsular Malaysia it is found in many places on the lowlands, particularly in fairly open places on trees or by the sea on rocks.

Cultivation

This species is quite popular among orchid growers. It bears rather attractive "bundles" of small scarlet flowers. It should be grown in a pot filled with rocks or charcoal and placed in the sun.

The reddish-orange flowers of *Renanthera matutina*. Note the two lateral sepals are held close together.

RENANTHERA

The name Renanthera is a combination of a Latin word *renes* = kidney, and the Greek word *anthera* or anther. Together they refer to the kidney shaped pollinia of the species. The name Renanthera was first used by Loureiro in 1790 to describe a native species of Indochina, *Renanthera coccinea*.

The flower of a Renanthera is very much similar to the Arachnis, the Scorpion orchid. To distinguish between the two, Holttum wrote that in the Renanthera "the lip is joined to the base of the column by a thin portion so that it is slightly movable, but there is no special hinge as in the Arachnis".

Three species of Renanthera have been listed by authors as native to Peninsular Malaysia. They are all well-known to most orchidists: *R. elongata, R. histrionica* and *R. matutina*.

With changes in nomenclature the first two species have been transferred to other genera: *Porphyrodesme elongata* (Bl.) Garay and *Renantherella histrionica* (Rchb.f) Ridl., respectively. So what is left of Renanthera that is native to Peninsular Malaysia is *R. matutina*.

Renanthera matutina (Bl.) Lindl.

Previous names:
Aerides matutina Bl.
Renanthera angustifolia Hk.f.

Stem: is long and climbing. Sometimes it grows hanging down.
Leaf: is thick and fleshy, rather stiff and greyish green.
Inflorescence: is long and little branched. The flower stalk is long and thick in relation to the sepal or petal.
Flower: Sepal is slightly broader than the petal but both are about the same length. Also both are long and narrow. The dorsal sepal and two lateral petals are arranged about 45° to each other. The two lateral sepals are close together but their tips diverge. The base of each of the lateral sepals is slightly twisted and this can be seen

fairly distinctly. The lip is small. The side-lobe is not distinct and appears as a small insignificant curved tip. The mid-lobe curves forward. A distinct line is seen dividing the lip structure into two equal halves. The flower has a long spur. Its apex is round. The column is short, thick and straight.

R. matutina is found in Java and Sumatra besides Peninsular Malaysia. Its habitat is the exposed rocky spaces of montane areas at an elevation of about 4,000 feet.

Cultivation

R.matutina is not a common Renanthera kept by orchidists. Growers prefer the more showy fire orchids of the Philippines: *R. philippinensis* and *R. storiei*. *R. matutina*, however, can be grown well tied to a post or in a pot.

The stalk of *Renantherella histrionica* is not twisted, hence the flower appears upside down.

4mm

RENANTHERELLA

This is a one-species genus named by Ridley as Renantherella. It is endemic to Peninsular Malaysia. The species, *Renantherella histrionica* is better known as *Renanthera histrionica* to most orchidists. In fact, Holttum treated it as a species of the Renanthera. To him, "the essential floral structure including the shape of the lip and top of the column are exactly typical of Renanthera". However, after close study, one could accept that there are enough differences between the two to warrant their separation. Ridley established a new genus for the species on the basis of the following differences:

a)   The leaf is semi-terete, succulent with a pointed tip. This is different from the Renanthera.
b)   The flower is inverted.
c)   The column is long.

Renantherella histrionica Ridl.

Previous name:
Renanthera histrionica Rchb.f.

Stem: is scrambling. It can grow quite long.
Leaf: is semi-terete, curved and fleshy. It has a pointed apex.
Inflorescence: is unbranched, horizontal and elongated gradually. One or two flowers open at a time.
Flower: The stalk is long. It is not twisted, hence, the position of the flower is "upside down", with its lip on the dorsal plane when it opens. The sepal is slightly broader than the two lateral sepals. All these lean straight backwards to about 45°. The two lateral sepals are shorter than the rest. They are close together and are below the lower plane of the flower. They curl backwards to form about half a circle. The petal and sepal are lemon-yellow with crimson spots near the edges and the tips.

   The lip is joined to the column-foot. The side lobe is broad, long, erect and leans forward. The mid-lobe is narrow and short. It curves forward forming about half a circle with the apex touching the spur. At the joint where the two side-lobes and the mid-lobe meet is a low callus formed by 4 ridges. The spur is very short, about the length of the mid-lobe. It is broad, tapering to a blunt apex.

The column is long and bends slightly. It is about $^3/_4$ the length of the dorsal sepal.

R. *histrionica* is endemic and is found throughout Peninsular Malaysia in areas from sea level to 4,000 ft. elevation. It grows in mangroves and on other trees or scrambling on rocks by the sea.

Cultivation

The plant is rarely cultivated, possibly because the inflorescence is too short, with few flowers. *R. histrionica* is easy to grow. It requires a pot filled with charcoal or broken bricks. The plant flowers rather frequently. The flower is quite attractive and unique.

A close-up of *Rhynchostylis retusa* showing its beak-like column.

3mm

RHYNCHOSTYLIS

This is a genus of about four species, consisting of *R. retusa, R. gigantea, R. coelestis* and *R. violacea.* Of these, the first two species are native to Peninsular Malaysia.

Rhynchostylis has a spectacular inflorescence of flowers. As such, the members of this genus are well-known among orchid collectors and growers. Locally, *R. retusa* is known as the foxtail orchid, while *R. gigantea* in Thailand is called the Elephant or "Chang" possibly due to the flower spike that resembles an elephant tusk. The generic name Rhynchostylis, is derived from the Greek words, *rhynchos* = beak, and *stylos* = pillar. This is in reference to the column of the flower. Indeed, a close look at the *R. retusa* flower reminds one of a dove (it's beak prominent) about to land in flight.

Rhynchostylis retusa Bl.

Stem: is short and thick. It is a monopodial epiphyte.
Leaf: is thick and leathery, rather long and narrow, unequally bilobed with several faint longitudinal lines.
Inflorescence: is drooping and unbranched. It bears numerous flowers close together. It gives an appearance of a long compact cluster, hence it is also called the "foxtail" orchid.
Flower: The sepal is larger than the petal. Both are white in colour with some purple spots. The lip is not hinged but joined to the column-foot. The spur is flattened and points backwards. The unlobed lip has a blunt, notched apex. It is this feature that gives the species its name "retusa".
The column is short and straight. The tip is beaked.

R. retusa is widely distributed. It occurs in Sri Lanka, India, Thailand, Indochina, Malaysia, Indonesia and the Philippines. The plant grows perched on trunks or branches of trees in exposed places. When in bloom, the inflorescence is a magnificent sight. In Peninsular Malaysia, the plant seems to be confined to the Northern States viz. Kelantan, Perak, Kedah and Perlis.

Rhynchostylis gigantea (Lindl.) Ridl.

Previous names:
Saccolabium giganteum Lindl.
Vanda densiflora Lindl.
Anota densiflora Schltr.

Stem: is short and thick
Leaf: is fleshy and leathery. It is unequally lobed at the tip and is variegated with light and dark green.
Inflorescence: is unbranched, arched or drooping. It can be as long as the leaf and can carry about 40 to 60 flowers. A large well-grown plant can produce many flower spikes.
Flower: The sepal is about twice as broad as the petal. It can be white or red depending on the plant type. The lip is joined to the column-foot. The spur points backwards and is parallel to the flower stalk. Its lip is pointed. The mouth of the spur is closed due to the lateral compression of the lip. The lip is straight. The apex is 3-lobed. The two lobes at the side are rounded. The middle lobe turns up slightly and is fleshy. On the underside of the middle lobe is a lump which gives it the appearance of a chin.
The column is short.

R. *gigantea* is found in Indochina and Thailand. Ridley said that the species is rare in Peninsular Malaysia though some specimens were found growing in the east coast of Johor and Singapore. Thailand seems to be the centre of development of this species. It is widely distributed, growing even up in the cool mountains of Chiengmai.

Cultivation

Both *R. retusa* and *R. gigantea* are easy to cultivate. The plant can be grown attached to tree trunks or branches or it can grow in an empty wooden basket. Alternatively it can be grown in a pot with charcoal or coarse broken bricks. The plant does well in shade with good drainage and aeration. In Thailand, it is well known as the Elephant orchid. It has been improved upon through hybridization with similar species and the products are made available for cultivation.

ROBIQUETIA

The genus Robiquetia consists of about 20 odd species found scattered throughout the South East Asian countries including New Guinea. One widely distributed species in Peninsular Malaysia is *R. spathulata*. Earlier on, many species of this genus were all grouped under Saccolabium.

Gaudichand originally established this genus in 1826. He named it Robiquetia in honour of a French chemist, Pierre Robiquet, who at that time had made many important discoveries, including caffeine and morphine.

Robiquetia spathulata (Bl.) J.J.S.

Previous names:
Cleisostoma spathulatum Bl.
Cleisostoma spicatum Lindl.
Saccolabium densiflorum Lindl.

Plant: resembles *Cleisostoma scortechinii* except that it is much larger. The stem can be long and climbing.
Leaf: is leathery.
Inflorescence: is single, pendulous and with numerous flowers arranged in a tapering bunch.
Flower: Stalk is about the same length as the spur. The sepal is slightly larger than the petal. Both are reddish-brown with yellow edges and midband. The spur is short, slightly flattened laterally and the tip blunt. There is no callus at the mouth of the spur as in the Cleisostoma; instead a small callus is present about $^2/_3$ way down the spur. It originates from the front-wall of the spur. About $^1/_3$ way down the back wall of the spur is a two-pronged, membranous tongue.

R. *spathulata* is found widely in the lowlands of Peninsular Malaysia growing on trees by rivers.

Robiquetia spathulata.

SMITINANDIA

This genus of two species was created by Holttum in 1969. It was named after Mr. Smitinand, a Thai who has contributed tremendously to the botanical knowledge of the orchids of Thailand.

Earlier, the species belonging to this genus were either placed in the genera Saccolabium or Ascocentrum, but careful examination of the fresh specimen by Holttum showed that *Smitinandia micranthum* has four pollinia while the Ascocentrum has two.

Smitinandia micrantha (Lindl.) Holtt.

Previous names:
Saccolabium micranthum Lindl.
Cleisostoma micranthum (Lindl.) King & Pantl.
Ascocentrum micranthum (Lindl.) Holtt.

Smitinandia micrantha, a rarely found species in Malaysia.

4mm

Stem: is short and erect.
Leaf: is fleshy. The tip is deeply rounded but unequally bilobed.
Inflorescence: is unbranched, rather long in relation to the stem. It is thick and fleshy.
Flower: The sepal is broad with a blunt tip. The petal is about the same length as the sepal but is only about half the width. The spur is about parallel to the ovary. The tip is tinted with purple. The midlobe is fleshy and oblong with a rounded end. The base of the midlobe is thick and forms an outgrowth to close the entrance of the spur. This characteristic can at first glance mislead one to consider this species as a Cleisostoma. The side-lobe is small, rounded and points forward. The column is short. On either side of the rostellum is a swollen area.

Smitinandia micrantha is distributed from regions of the Sikkim Himalaya down through Tenasserim in Burma, Laos, Cambodia, Thailand up to Langkawi and Kedah, the northern states of Peninsular Malaysia.

Cultivation
The plant is not common. It can be grown tied to a post in the shade.

Smitinandia micrantha with its thick and fleshy inflorescence.

SPATHOGLOTTIS

This is a genus consisting of about 40 species. They can be found in regions from India, southern Japan and China down through the Malay Archipelago, the islands of the Pacific and northern Australia. Six species are native to Peninsular Malaysia. The commonest and most widely distributed is *S. plicata*. The curious little species *S. hardingiana* grows on the limestone areas of Langkawi Island, having a similar habitat as *Paph. niveum*. This species does not seem to be worth cultivating. The four other species are highland plants. They bear golden yellow flowers which are much loved by growers.

The generic name Spathoglottis was given by Blume and is derived from the Greek words *spathe* = a broad blade, *glossa/glotta* = tongue. Obviously, this refers to the characteristic broad, spatula or spoon-like lip of the flower.

Spathoglottis plicata Bl.

The lip of *Spathoglottis plicata*.

8mm

Pseudobulb: is ovoid and grows in clusters.
Leaf: is lanceolate and can be broad and long. Its specific name, *plicata* refers to its pleated or narrow folded leaf blade. A few leaves arise from each pseudobulb.
Inflorescence: is long and arises from the base of the pseudobulb. Flowers are borne in a cluster at the end of the spike. Two or three flowers open at a time. As such the plant remains in bloom over a long period of time.
Flower: The sepals are slightly narrower than the two lateral petals. They can be of different shades of purple, mauve or white. The lip is distinctive. The side-lobes are erect and they bend over to enclose the column. The mid-lobe is narrow in the middle but flares out at the tip to give the shape of a broad tea-spoon or spatula. On either side at the base of the side-lobe is a callus. This can be bright yellow in colour. Below each of the calli is a little hairy structure pointing vertically downwards.
The column is long, narrow and winged. Its tip bends inwards.

The white form of *Spathoglottis plicata*.

S. plicata is a hardy, extremely variable species found growing in open fields, along streams or on edges of rock either in the lowlands or elevated areas. The species is very widely distributed.

Spathoglottis affinis de Vr.

Pseudobulb: is small and flattened and is irregular in shape.
Leaf: is narrow, long and pleated.
Inflorescence: is long, slender, and erect. It is covered with fine, short hairs. The inflorescence has many flowers but only two or three open at a time. The flower stalk is slender, greenish in colour and is covered with fine white hairs.
Flower: The dorsal sepal is slightly narrower than the two lateral ones. The sepals are covered with fine short hairs while the petals are not. The dorsal sepal and two lateral petals are entirely golden yellow in colour. The two lateral sepals are also yellow but have three reddish brown streaks on each of them. The side-lobes are erect. At the base of each side-lobe is an upright bright yellow callus with reddish brown spots. On the underside of each callus is a lateral appendage and this is covered with fine white hairs. This structure is entirely yellow in colour. The mid-lobe is narrow in the middle but broadens out towards the tip and is notched.
The column is long, broadly winged and is about the same length as the side-lobe. It arches over to about half a circle.

Details of the lip of *Spathoglottis affinis*. Note the yellow appendage on the underside of the callus.

The delicate inflorescence of *Spathoglottis affinis*.

S. affinis is a mountain plant. It is native to the montane areas of North Peninsular Malaysia. There is a similar species called *S. lobbii* which inhabits the highlands of Thailand, Indochina and Myanmar. Both these species appear similar in description.

S. affinis can be grown and will flower in the lowlands. However, it appears fragile and a weakling. Indeed, this golden Spathoglottis is worth cultivating! There is one point which each grower needs to bear in mind. The plant is deciduous and loses its leaves during the dry season. During this time the tuber needs rest and during this period it needs to be given the minimum of water and fertilizer. When new growth begins, it should be exposed to full sunshine to enable good flowering. The plant can remain in flower for three to four months while it remains more or less leafless.

Spathoglottis aurea Lindl.

Previous name:
Spathoglottis wrayi Hk.f.

Spathoglottis aurea, a montane species.

Pseudobulb: is ovoid.
Leaf: is lanceolate and broad.
Inflorescence: is long and thick as in *S. plicata*. The flowers that are borne at the end of the spike are deep yellow in colour.
Flower: The sepal and petal are about the same size as each other. The side-lobes are flushed or spotted with crimson. The base of the mid-lobe and the calli have small crimson spots or streaks. The mid-lobe is rather narrow and does not expand at the tip to form the spatula-shape which is characteristic of the genus. The column is arching.

S. *aurea* is found in Java and Sumatra besides Peninsular Malaysia. It is a montane species found growing on grassy spots at 3,000 to 6,000 ft. elevation.

Cultivation
Spathoglottis seems to be the most commonly grown ground orchid. Being a terrestrial it is generally treated like any ornamental plant. It is grown in a pot or in the ground, placed either in partial shade or full sunshine. The plant should be watered frequently. Organic matter or compost should be added once in a while. It is a hardy plant and can remain in bloom almost throughout the year if well taken care of.

THECOSTELE

Allied to the Acriopsis, the genus Thecostele is similarly small, comprising about 5 to 6 species. Its habitat ranges from North East India through to Malaysia and the Philippines. Two species are native to Peninsular Malaysia: *T. alata* and *T. secunda.* Another species, *T. manqayi* or *T. quinquefida,* was recorded in Hooker's Icones Plantarum as having been found in Melaka, but according to Holttum this could be just another species of *T. secunda.*

Thecostele alata (Roxb.) Par et Rchb.f.

Previous names:
Cymbidium alatum Roxb.
Thecostele zollingeri Rchb.f.
Collabium wrayi Hk.f.
Thecostele maculosa Ridl.

Pseudobulb: is large, flattened with a few ridges.
Leaf: is long and broad. There is only one leaf on each pseudobulb.
Inflorescence: is unbranched, long and pendulous. It carries many well-spaced flowers.
Flower: Each flower is rather small, stiff or leathery in texture. The sepal is wide and keeled with its terminal end curving forward. The two lateral petals are narrow and the tip curves forward like the sepal. The sepal and petal are pale yellow at the base, white in the middle and light purple at the tip. There are irregular dark purple spots on them.

There is a tube-like structure or nectary attached horizontally and parallel to the underside of the foot-column. The lip is hinged to the lower rim of the opening of the tube.

The sidelobes are erect and form two curved arms. The foot-column fits tightly in between these arms. The mid-lobe is broad and has a gentle notch. It is white with a purple tip. The centre has large purple markings. Placed between the side-lobe and the mid-lobe are two short keels. The entire surface of the lip is covered

Thecostele alata. Note the column which arches to form a semi-circle and the two side-arms at the end.

0.5mm

with coarse hair-like or papillae-like structures.

The column bends over to form about half a circle. Near the tip are two curved side arms. These arms seem to act as a guide to align the pollinating agent to come into contact with the pollinia and the stigma when visiting the flower. Given a floral structure such as this, the pollinating agent possibly needs some skill to get to the nectary. It has to forcefully tilt the tip of the mid-lobe to release the "lock" formed by the side-lobes and the foot-column. When the lip is released it swings over and the opening of the nectary becomes accessible. Taking the flower structure as it is, the nectary opening remains effectively closed.

Thecostele alata can be found in many localities in the lowlands. The plant, however, is not common.

Cultivation

The plant is not grown by hobbyists and like many others is only of botanical interest. It grows well under partial shade tied to a wood-piece or tree trunk.

THRIXSPERMUM

The genus Thrixspermum possibly needs re-study. There are various opinions as regards which species ought to rightly belong to this genus. Ridley listed 10 species under Thrixspermum that are native to Peninsular Malaysia. Other "similar" species were listed under the related genus Dendrocolla. Holttum in his book, *Flora of Malaya: Orchids,* grouped about 35 species under Thrixspermum. He, however, recognised the need to divide the genus into two sections: Orisidice and Dendrocolla. To some authors, these sections are in fact separate genera by themselves. Some members of the Thrixspermum were also included under Sarcochilus by some people.

The name of this genus is derived from Greek, *thrix* = hair, and *sperma* = seed. This is in reference to the long, thin seeds. The flowers of Thrixspermum are rather fragile and short-lived. In some species the flower opens in the morning and closes in the evening of the same day. In some species, the flowers on the inflorescence are produced in succession over a period of time.

Members of Thrixspermum are distributed from Ceylon to the Pacific Islands including Japan. It is well represented in Malaysia, the Philippines and Indonesia.

Thrixspermum calceolus (Lindl.) Rchb.f.

Previous name:
Sarcochilus calceolus Lindl.

Stem: is long and climbing.
Leaf: is unequally bilobed, and the apex is rounded. There are many leaves on each stem.
Inflorescence: is short, bearing one to three flower buds.
Flowers: open singly. They are fleshy, white and scented but delicate and short-lived. The dorsal sepal is slightly narrower than the two lateral sepals. The two lateral sepals are grooved in the middle. The two petals are narrower than the dorsal sepal. The lip is fixed to

The fleshy, solid mid-lobe of *Thrixspermum calceolus.*

7mm

The white, fleshy flower of *Thrixspermum calceolus*.

the foot-column. It is immovable and brittle. The mid-lobe is fleshy and forms a solid tongue. It has a yellow band. A sac, which is characteristic of this genus, is present at the inner end of the mid-lobe. At the front wall of the sac is a prominent white callus. At the base of the sac is a yellow patch with numerous brown spots. The column is short.

Thrixspermum calceolus is found growing in clumps in the open or slightly shady areas of the lowlands. It grows on trees or rocks throughout Peninsular Malaysia. The species also occurs in Indo-China and Indonesia.

Cultivation

The species is not cultivated. Though the flower is attractive it is very short-lived. The plant grows easily, tied to a wooden support and placed in light shade.

TRICHOGLOTTIS

This genus has about 60 members distributed throughout Eastern Asia, Polynesia and the adjacent islands. It is particularly well represented in the Philippines with 16 species, of which *T. brachiata* is very attractive and unusually coloured.

Within this genus there seems to be two vegetatively different groups of plants. One type, as exemplified by *T. cirrhifera,* is a climbing plant. In nature it grows hanging down. The flower is small. The other type, as in *T. brachiata* and *T. fasciata,* has large flowers and the plant grows upright.

The name Trichoglottis is of Greek origin, *thrix* = hair, and *glotta* = tongue, referring to the presence of hairs in the throat of the lip. There are two characteristic features of the genus:
1) The lip, especially the mid-lobe, is rather complicated and it can be hairy.
2) There are horns on the column.

Trichoglottis cirrhifera T. et B.

Previous names:
T. tetraceras Ridl.
Saccolabium cornigerum Ridl.
Cleisostoma tenuicaule King & Patl.

Stem: is long, grows hanging down.
Leaf: is spreading, narrow with pointed apex.
Inflorescence: is uni-flowered, with one or two at each leaf axis.
Flower: has a white stalk about twice the length of the spur. The sepal is slightly broader than the petal. Both have rounded apices. They are yellow with broad brown blotches. This brings to prominence the lip which is white in colour.
The column-foot is joined to the horizontal spur and at the side of this joint is attached the lateral sepal. The mid-lobe is white and is shaped like a tongue. It is not hairy. The side-lobe forms a pointed

Trichoglottis cirrhifera with the side lobes forming 'V'-shaped arms.

arm. Between the two arms (side-lobes) is a furrow. This is an extension of the mouth of the spur. The tip of the furrow has a violet marking. The furrow continues behind to form a tube, or the spur. The spur is white, horizontal and placed close to the flower stalk. The apex is rounded.

The column is short, broad, and has two horns protruding from the two frontal corners of the column which is yellow in colour.

T. cirrhifera is found in the northern states of Peninsular Malaysia. This is a variable species.

Trichoglottis fasciata Rchb.f.

Previous names:
Stauropsis fasciata Benth
Stauropsis fasciatus Ridl.

Stem: is erect-growing.
Leaf: is large, measuring about 12 cm x 2.5 cm and its tip unequally rounded.
Inflorescence: is rather long compared to other species in the genus. It has 2 to 4 flowers which are also large. The inflorescence stalk is angled with two wings above, forming a shallow furrow.

Flower: The sepal is broader and slightly longer than the petal. Both are keeled at the back towards the tip. The tips of the sepal and petal are broad but terminate in a pointed apex. The sepal and petal are lemon yellow with transverse brown bands. The lip is not spurred and is white with some brown spots. The side-lobes are erect, held parallel and close to one another. This forms a deep furrow about half the length of the lip. The mid-lobe has two flat, lateral arms placed at right angles on either side at about the mid-length of the lip. This arm is horizontal and is triangular in shape. Its surface is covered with numerous fine hairs. The apex of the mid-lobe forms a single, flattened, vertical structure. Its upper surface is hairy.
The column is short and broad. Long horns protrude out of the frontal edge of the column.

T. *fasciata* is found in Thailand and the distribution extends to the northern regions of Peninsular Malaysia. The species is also found in the Philippines and Sumatra.

Trichoglottis fasciata.

Trichoglottis lanceolaria Bl.

Stem: is pendulous and slender. The roots cling onto small branches or twigs. Plant is small.
Leaf: is narrow and pointed.
Inflorescence: consists of one or two flowers borne at the node.
Flower: Flower stalk is tri-angled and is lightly grooved. It is yellow in colour, slender and about $1^{1}/_{4}$ times the length of the dorsal sepal. The dorsal sepal and the two lateral petals are about the same size and length as one another. These three structures are not spreading but point forward forming a three-sided hood over the column. The lateral sepal is broad and is about triangular in shape. Its lower side is slightly broader with a pointed corner. These lateral sepals form the "mentum" hugging the spur. The sepals and petals are all light yellow in colour. The mid-lobe of the lip is whitish yellow, broad and translucent with fine papillae around the opening of the spur. The apex is blunt with a notch in the middle. About three-fourths of the way down the middle of the lip is a deep hollow. The side-lobe is squarish and bends forward. It is upright like the collar of a shirt and has streaks of reddish purple. The side-lobes form a circle over the spur. At the mouth of the spur is a purple colouration. A yellow tongue arises from the back wall of the spur at about the same level as the opening of the spur. The spur tapers to a blunt apex, lemon yellow in colour and is partitioned by a thin membrane vertically. The spur hangs down parallel to the lip. They are attached at right angles to each other. The front wall of the spur is thick and in the middle is covered with fine hairs. The column is about $^{1}/_{4}$ length of the dorsal sepal and is broad.

Trichoglottis lanceolaria as described here differs from what has been written by Holttum.
The following are a few discrepancies:
(1) The colouration does not match Holttum's description.
(2) The shape of the side-lobe of this specimen is squarish, not triangular.
(3) There is a deep hollow in the mid-lobe, and this could possibly match Holttum's description: "...at the base of mid-lobe a transverse deep yellow band and two low hairy calli which diverge to meet the front edges of the side-lobes".

Trichoglottis misera (Ridl.) Holtt.

Previous name:
Saccolabium miserum Ridl.

Stem: is long-climbing.
Leaf: is long and narrow, rather leathery.
Inflorescence: is single-flowered and several can form at a node.
Flower: The sepal is longer and slightly broader than the petal. The tips of the sepal and petal are broad and flat. Both are pale yellow in colour. The lip is white in colour. The mid-lobe is broad and elliptical. The surface is covered with fine papillae. The throat

The solitary inflorescence of *Trichoglottis misera*.

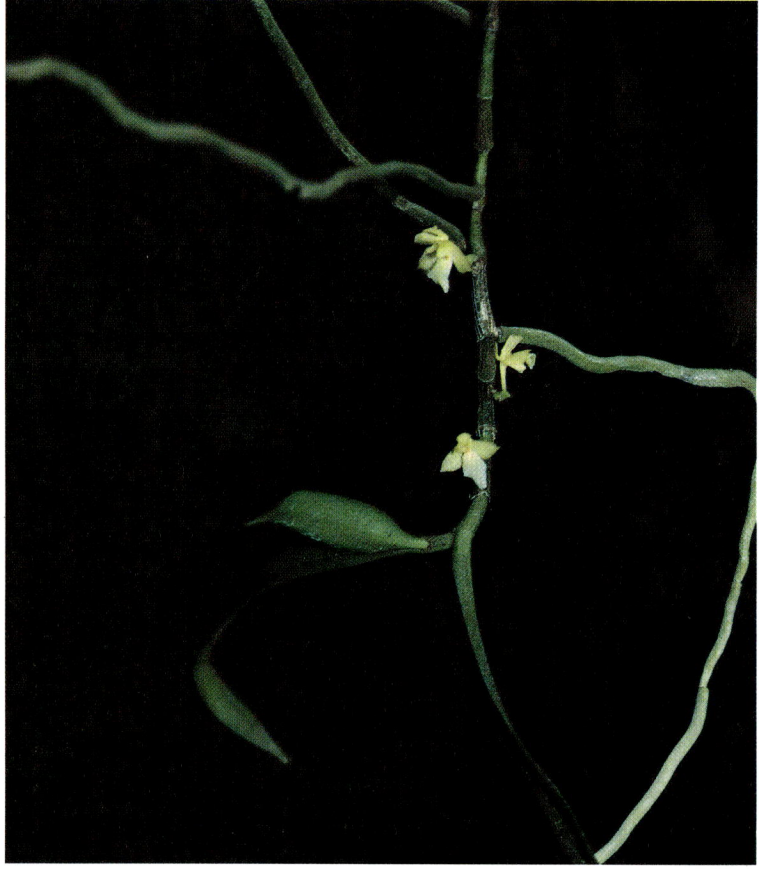

A cross-section of *Trichoglottis misera* showing the shape of the spur.

of the lip has purple markings.In front of the opening of the spur is a callus with three gentle ridges.The side-lobe is erect and rectangular in shape. The spur is about $1/4$ the length of the mid-lobe It is broad but then tapers to a point.

The column is short and broad. The horn on the column is not prominent. This is one species in which the typical horn of the genus is absent. Its tongue, though, is typical of the Trichoglottis.

T. misera is found in the northern part of Peninsular Malaysia.

Trichoglottis retusa Bl.

Stem: is long-climbing. Usually it grows hanging down.
Leaf: is oblong with its apex unequally bilobed.
Inflorescence: is single flowered.
Flower: The sepal is slightly longer than the petal. Both are reddish-brown, and pale greenish yellow at the edge. The lip has a short spur and is of a lighter colour. The mid-lobe is tongue-shaped, narrow and short with hairy surface. The side-lobe is narrow, curved outwards and upwards. At the base of the side-lobe is a small, hairy yellow callus.

The column is short with a hairy horn.

T. retusa is found growing on trees and limestone rocks in Central Peninsular Malaysia. The species can also be found in Indochina, Thailand and Indonesia.

Cultivation

Most Trichoglottis are generally not cultivated by orchid growers, with the exception of *T. brachiata* which has a very deep chocolate colour and is unique. The Malaysian species *T. fasciata* is also attractive and is an item of interest to collectors. For the two species above, the plant should be grown in a pot or in the ground and provided with a support. The roots need good moisture and are not hardy so it should be grown in the shade.

VANDOPSIS

Vandopsis is a genus of about 8 species native to regions in Myanmar, South China and islands in the Malay Archipelago. One species, *V. gigantea* is native to Peninsular Malaysia.

Vandopsis is a genus established by Pfitzer in 1889. The name means Vanda-like. The word *opsis* is Greek for resemblance.

Vandopsis gigantea (Lindl.) Pfitz.

Previous names:
Vanda gigantea Lindl.
Stauropsis gigantea Benth.

Stem: is stout and rather large.
Leaf: is fleshy, thick and the tip rounded but unequally bilobed.
Inflorescence: carries about a dozen flowers with a thick stalk.
Flower: sepals are slightly broader than petals. Each sepal is broadest in the middle and tapers towards both ends. Both sepals and petals are of very thick texture, dull yellow with red-brown blotches. The lip is firmly fixed to the base of the column by the side lobes. The mid-lobe is long, narrow and solid with a mid-ridge at the top. About $3/4$ length inwards, the ridge is interrupted by a V-shaped notch, and then the ridge continues. At the end of the ridge is a small protrusion on each side. The side-lobe is vertical and has a flap on its inner side. These flaps join to form V-shaped walls. The short, stout column sits above this V-shaped structure. Between the column and the V-shaped side-lobe flap is a gap which forms a cavity. On the underside of the column is the large, deep hollow of the stigma.

Vandopsis gigantea is distributed from Myanmar down through Thailand and the northern half of Peninsular Malaysia. It is found growing on trees and rocks by the sea.

VANILLA

The world knows of Vanilla as a flavouring essence used in ice-cream more than as an orchid with fragrant flowers. This vanilla of commerce is *V. planifolia,* a native of Tropical America. It is being grown on a large commercial scale in Mexico and Africa for its fruits or pods. The pod has to be fermented before the vanillin can be extracted from it.

There are about 65 species of Vanilla found throughout the tropics and 5 species are native to Peninsular Malaysia. None of the Malaysian species can be grown for its vanillin though the fruit is said to be sweet and edible. When the plant blooms, and this occurs only rarely, the flowers emit a sweet aroma paralleled only by the world's best perfumes! Its flowering is therefore always looked forward to.

The name vanilla originated from the Spanish *vayailla,* meaning "little sheath or pod". This is an obvious reference to the long, slender fruit or pod produced by the plant.

Vanilla vine and seed pod.

Vanilla griffithii Rchb.f.

It has the same vegetative habit as *V. pilifera*.
Inflorescence: can carry 10 to 12 flowers.
Flower: The stalk is round and thick. The sepal and petal are about the same size as each other. Both are spreading and white with pale green flushes. The outer side of the lateral petal has a prominent mid-ridge. The lip is broad. It is white with a pale yellow tip. The base of the lip is flat, ovoid in shape and has red-brown bands. From this base, the side-lobe arises perpendicularly. It is broad with a wavy edge. The side-lobe is joined to about $1/4$ length of the column. At the front portion of the flat base is a round structure. Placed perpendicular to this ball is the mid-lobe, a large semicircular structure. The outer portion of the lobe is thin and has frilled edges. Inwards it is thick and becomes a U-shaped callus with hairy projections. The spherical structure and the U-shaped callus fit closely into each other.

The column is thin and is about half the length of the petal or sepal.

V. griffithii is common in Peninsular Malaysia and found scrambling on trees of the lowland forests. It frequents fairly open spaces. Though the species has not been reported to occur in Thailand, it is most likely that the distribution of this species could extend northwards into Peninsular Thailand.

Vanilla pilifera, a sweet smelling orchid.

Vanilla pilifera Holtt.

Stem: is a vine, with roots forming at each node. The roots support the plant as it climbs the host.
Leaf: is large and succulent. It narrows towards the top.
Inflorescence: is short with a few flowers.
Flower: is large with round thick stalk. They bloom one or two at a time. The sepal and petal are about the same size. The sepal is pale green. The petal is cream with a green keel on the back. The lip is cream with pale pink veins. It is thin and forms a "tube" with the side-lobes joined to the straight column. In this species, $3/4$ of the length of the column is joined to the lip. This is one of the characteristics which distinguishes the species from *V. griffithii*. In *V. griffithii* the lip is joined to about $1/4$ the length of the column. At the tip of the mid-lobe is a cluster of thick hair-like projections. About a quarter way inside, on the same alignment, is another group of thin broad-toothed structures filed closely together. At first glance this structure appears to be a solid column. The site of this structure is about the same position as the stigma above. From this it appears that this structure could play a role in the pollination of the flower. The constriction of the "flower tube" forces the insect to squeeze through it, ensuring that the back of the insect rubs against the stigma. The column is long and bends slightly at about $1/4$ its length from the tip. Otherwise, the column is straight and appears as a continuation of the flower stalk.

V. pilifera is found in Thailand as well as Johor, the southern tip of Peninsular Malaysia.

Cultivation
Vanilla is not cultivated by orchid growers. Being a climber, the plant is not suitable for the pot, besides it rarely blooms. It could be a good specimen for large gardens where the plant can be allowed to climb onto trees.

VENTRICULARIA

This is a new genus established by Garay. The genus is characterised by the belly-shaped spur of the lip. The word *venticulus,* from which the name of the genus is derived, means "belly". The type species of this genus is *V. tenuicaulis* (Hook.f.) Garay. Earlier, Holttum classified the species as Uncifera, a genus of a few species native to Myanmar and the neighbouring countries. Holttum, however, pointed out that the Malayan species differs in some ways from the Burmese species. Perhaps for this reason, Garay found it fit to establish a new genus for this species.

Ventricularia tenuicaulis (Hook.f.) Garay.

Previous names:
Saccolabium tenuicaule Hook.f.
Uncifera tenuicaulis (Hook.f.) Holtt.

The plant may grow into large clumps. At the outset, it looks like *Trichoglottis lanceolaria* and is easily mistaken for it unless a close examination of the flower is made.

Stem: is slender and pendulous.
Leaf: is narrow, gradually tapering to a sharp tip.
Inflorescence: is short with one or two flowers.
Flower: The sepals and petals are yellow in colour. They are elliptical in shape and are concave. The dorsal sepal appears like a hood to cover the column. The lip is white and is spurred. The spur is round-bottomed with a constriction about mid-way.
A cross-section of the spur shows the presence of numerous hairs on the front wall at the point of the narrow constriction.

V. tenuicaulis grows on thumb-sized twigs hanging down the canopy of trees. In Penang it is found in the lowlands and the seashore. The plant is also found on limestone areas.

The flowers of *Ventricularia tenuicaulis*. The one on the left has a sepal removed to show the spur.

0.5mm

Cultivation

The plant is not cultivated at all. It looks petite and hangs rather gracefully. The flower however is too small to be appreciated by anyone but serious collectors. Ventricularia should be grown tied to small posts and placed in the shade.

CULTIVATION OF SPECIES ORCHIDS

The cultivation of species orchids in Malaysia is not popular. This could be for two reasons. First, few species bear large, eye-catching flowers. As such, to laymen they are unattractive. Secondly, growing species may be difficult for the novice. They lack the vigour necessary for growth and may demand more precise conditions to grow well. As discussed in the earlier chapter, each species in its native habitat survives due to the presence of certain unique growing conditions. Having a large number of species in the garden "under one roof" therefore may not be adequate to provide all the different environmental niches required by each species.

Possibly because of the above reasons many orchid growers are more inclined to cultivate hybrids. Indeed they may well be justified in doing so. Nevertheless, to those who want to prove they have "green fingers", growing species is a challenge. In addition, the variation and uniqueness one finds in the flowers of various genera ought to be reason enough to make the hobby worthwhile.

There are three aspects to consider when growing species:
(a) There is a need for careful observation of the plant in its habitat. This provides clues to its optimum growth requirements.
(b) The plants collected from the wild have to be established.
(c) The established plants have to be properly maintained.

Habitat Observation

The careful observation of the growing conditions of the plant in its native habitat is important. This greatly facilitates creation of a favourable environment for its subsequent cultivation. If such observation cannot be made firsthand, then the grower must resort to the available literature or the experience of others. Depending on literature alone though, could have some drawbacks. Also, book identification is a difficult and laborious task. Hand a pile of orchid books and an unidentified flower to the "unseasoned" and see how he gets along! He would probably end up being "allergic" to orchids all his life! As much as this sounds pessimistic, one should not be dismayed. Try and consult others who are familiar with the species. Experience is an excellent teacher and once the ability to grow a few species has been acquired, the proficiency follows on to other species.

Take careful notes while on collecting trips. Parameters such as altitude, seasonal variations, rainfall pattern, etc. are important. Generally, species collected from the cool altitudes do not grow or flower in the hot lowlands. If it grows at all in the lowlands, the plant may be weak, growth retarded and rarely with flowers. When growing orchids in higher altitudes, they are best placed in the most sheltered and coolest part of the garden or orchid house. Seasonal variation of the climate is an important factor to note. For example, when one is collecting plants from the northern parts of Peninsular Malaysia, it should be noted that there is a dry spell from January to March each year. Though not all orchids require this dry spell for growth, there are, however, many species which depend on this dry spell to stimulate them to flower. The terrestrial orchids in particular need this dry spell and this constitutes the "resting" stage in their growth cycle. A grower should subject his plants to similar conditions.

Orchids do not grow everywhere. A particular orchid has its favourite niche among the vast range of habitats. For example, it is very rare to see *Dendrobium crumenatum* on the underside of a branch, while many Cleisostoma species always find their home there. The size of the branch is also important. *Dendrobium crumenatum* prefers a large branch but since the underside of such a branch is rather dry, one does not normally find them there. In comparison, Cleisostoma sp. grows on small branches overhanging streams. In this environment, the underside of the branch is more humid than its upper side, hence allowing the plant to grow luxuriantly.

One can generalize that the smaller the host branch the more light the species growing on it will accept.

The water retention capacity of the host need also be considered. A root that grows in moss or the forest litter needs a damper environment to survive in than one which grows on exposed bark. The danger of drying out for a plant which has been growing on exposed bark is much less than for some terrestrials when they are brought home for cultivation.

One other important point to note is the growing medium. Some species grow in limestone areas while others are found only in the deep leaf litter of the forest floor. Yet others grow on one particular host tree only. All these points determine their subsequent cultivation and could often mean the difference between success and failure.

Establishment

To go collecting orchids in the jungle is an activity by itself and to come home and try establishing them in the garden is yet another. It is fun trekking the forest, but often greed wins the day and one grabs as many plants as one can find. After all Mother Nature provides them for free. Unfortunately when one returns home, the plants are left stranded and unattended. The fun is all over. Such a trait of destruction and abuse must not be allowed to exist in the heart and soul of an orchid lover!

The period between collection and when the plant is firmly attached to the host tree in the home garden is important and requires close attention. The plant can be tied on a tree or grown in a pot. Irrespective of what is done, there is a need for careful observation to see if the plant takes to the new home. If the siting of the plant in the garden is not correct then move it to another location. If the medium provided is unsatisfactory, change it for a better one. If more than one plant is collected, they are best grown on different media or sited at different locations. When one plant has shown signs that it prefers a particular locality then shift the rest to the one just established. Generally, a plant should establish itself within six months. Failure to do so warrants re-examination to improve technique.

When collecting the plant it is important to retain as many roots as possible. Also any moss or litter that are associated with them should be retained. The plant should not be allowed to dry out or be exposed to the sun. At home they should be planted as soon as possible.

Roots that are badly damaged should be trimmed off, but in cases where most roots are damaged even some of the damaged ones should be retained.

One can grow species on either artificial or natural hosts. There are advantages associated with both of these methods. Orchids can be grown on crushed cork slabs of about one inch thickness. This medium is popular for growing Bulbophyllum, Sarcanthinae, and the smaller Dendrobium species. Unfortunately cork slabs are not readily available in Malaysia. The advantages of using this medium are:

1) It has long life
2) Its pH is neutral, i.e. it has neither acid nor basic reaction as a growing medium.
3) It scarcely retains salts
4) The slab provides wet surfaces for the roots to penetrate into while it remains dry outside.

In Malaysia, a substitute for cork slabs could be fern root slabs and portions of cut branches. Coconut husks can also be used successfully and it is cheap too. There are, however, some drawbacks in using it. It may retain too much water or encourage excessive growth of mossed or algae on it. Growth of these organisms is undesirable from the aesthetic and physiological point of view. Functionally they may affect the oxygen supply to the orchid roots and also may alter the pH of the medium. You can remove the mosses or algae from the roots by gently brushing them off with a toothbrush. They come off easily. Root vigour is restored after doing this. Another disadvantage of coconut husk as a medium is that it breaks down after 3 to 4 years.

Growing plants in a pot is satisfactory for many species, especially those with penetrating roots, e.g. Cymbidium. A suitable potting medium is a mixture of charcoal and broken bricks. For the non-deciduous terrestrial orchids, leaf mould and humus is very suitable. The deciduous terrestrials, however, require an open medium which drains well.

Indeed, growing deciduous terrestrials can be an art in itself. Any prospective grower of such plants would find it worthwhile to consult Australian orchid journals as deciduous terrestrials are widely grown there. Many years have been spent in trying to perfect the growing technique.

Cultivating species orchids on an artificial medium provides flexibility. This is an advantage because the plant can be easily transported from one place to another.

Growing species on natural hosts provides less flexibility. In addition, the garden has to have suitable host tress or one has to plant some. They could take some time to establish themselves. The author prefers this technique as it provides a more natural setting and adds beauty to the garden. The Penang Botanical Gardens and the newly established Orchidarium of Universiti Sains Malaysia at Muka Head adopted this approach of planting species orchids.

To be successful at planting orchids this way demands familiarity with the growing conditions of the species in the wild. The aim then is to duplicate the conditions that are necessary for growth in the new environment. The advantage of this approach is it requires less maintenance once the plants have been well and satisfactorily established. Hopefully then, the plants can perpetuate themselves in the years ahead.

Common trees in the gardens, such as rambutan (N*ephelium lappaceum*), mangosteen (*Garcinia mangostana*) and frangipani (*Plumeria sp.*) prove suitable hosts to many species. The author has seen a rambutan tree in the garden of a friend which is a host to numerous different species. All of them grow well and happily as one big family under one roof! There is little use climbing to the top of the tree to plant orchids as a similar environment can be found amidst the lower canopy of the tree. Keep the orchids on branches as low as possible so that one can appreciate their flowers without having to strain the neck too often, or use a ladder.

For an isolated tree, it must be remembered that the western side of a tree is drier than its eastern side. This is due to the drying effects of the afternoon sun. Knowing this, select the locality for the orchid accordingly.

The smooth shiny bark of the frangipani makes a dryer host than the rambutan bark. In contrast, the mangosteen bark is smooth but it retains moisture. It is one of the preferred hosts for orchids. In Penang, *Kingidium deliciosum* is commonly found on mangosteen trees.

The value of trees as host for orchids is indicated by the presence of lichens or mosses growing on them. A host that supports these would invariably support orchids as well.

When growing a plant, ensure that it is securely tied onto the host. Use a nylon fishing line. The common string does not last and can break off before the orchid is fully established. The use of ordinary wire may damage the plant or the host if it is not removed after some time. Also, wire may rust and this is not desirable.

When planting terrestrials in the ground, ensure that they are not planted too deep, no deeper than how they are generally found in nature. For example, with Bromheadia and Arundina, their crowns should be just below the soil surface, while with Phaius and Calanthe ensure that the roots are well buried in the leaf litter.

Due to the loss of roots during collection, the newly planted orchids will need extra watering initially. Keeping in mind that dryness stimulates root growth, the plants must be allowed to dry out in between watering. Never allow it to remain damp throughout the day.

Maintenance

Once established, the plants can be treated in various ways to ensure flowering and proliferation. Flowering could be stimulated by climatic factors such as light intensity, a sudden drop of temperature, a dry spell, or wetness after a dry spell.

Sunlight hours are important. Plants should be progressively given brighter conditions to promote better growth and flowering. A drop of temperature occurs as a natural phenomenon and therefore there is nothing much a grower can do to achieve this effect. Conditions of dryness are just as difficult to create for plants growing on a natural host. Some species may require some months of dryness for completion of their growth cycle. During this "rest" period watering has to be restricted to only once every week or 10 days.

Species orchids do not have the branching vigour of hybrids. Branching can be induced by cutting back the plant in the case of monopodials or division of the pseudobulb for the sympodials. When dividing the pseudobulbs, cut out a group of at least 3 to 4 pseudobulbs. If a division has less than this number, the new growth may lack vigour. Consequently, it may be stunted and flower poorly.

Application of fertilizers and pesticides to the established plant is necessary. There is no hard and fast rule to follow in fertilizer application. There are however, a few points to note:

1) Foliar application is preferable. In this case the fertilizer solution is sprayed to cover the host and plants alike.
2) Generally use a dilute dosage. Over-concentrated fertilizer solution can cause scorching of the leaves or even kill the plants outright.
3) Plants require nutrients most during active growth. So provide them at appropriate time and frequency.
4) Caterpillars and other pests can damage flower spikes and new shoots. Regular spraying with a suitable pesticide is helpful and sometimes necessary.

INDEX

Page numbers in **bold** denote on which page illustrations, or illustrations and related text, appear.

Acriopsis 7-8
Acriopsis javanica 7, **8**
Acriopsis ridleyi **8**
Aerides 9-10
Aerides lawrenceae 10
Aerides odorata **9**
Aerides odorata var. *bicuspidata* 10
Aerides odorata var. *cornata* 10
Aerides odorata var. *immaculata* 10
Aerides quinquevulnerum 10
Aerides odorata var. *suavissmun* 10
Aerides odorata var. *virens* 10
altitude 113
anggerik bulan 70
Anoetochilus 5, 11-13
Anoectochilus albolineatus 11, **12**, 13
Anoectochilus geniculatus 13
Anoectochilus pomrangianus 11
Anoectochilus reinwardtii 13
Anoectochilus repens 11
Anoectochilus tonkinensis 11
aphyllorchis 2
Appendicula 14-17
Appendicula cornuta 14, **15**
Appendicula pendula 15, **16**
Appendicula undulata **17**
Arundina 1, 2, 20-21, 60, 116
Arundina graminifolia **20**
Ascocentrum 18-19
Ascocentrum miniatum 18, **19**

bamboo orchid 21
Blume 50, 52, 70, 92
bromheadia 2, 116
Bromheadia finlaysoniana 5
Bukit Yong 67
Bulbophyllum 3, 5, 6, 22-31
Bulbophyllum biflorum **23**
Bulbophyllum corolliferum **24**, 31

Bulbophyllum lilacinum 24, 25
Bulbophyllum lobbii **25**, 26
Bulbophyllum longiflorum **26**, 27
Bulbophyllum maximum 6, **28, 29**, 31
Bulbophyllum medusae **30**
Bulbophyllum patens 30
Bulbophyllum sessile 30, 31
butterfly orchid 70

calanthe 2, 116
Calanthe vestita 3
callus 39
Cameron Highlands 3, 11
Cheang Kok Choy 32, 59
Chiloschista 32-33
Chiloschista sweelimii 22, **33**
Chiloschista usneiodes 32
cirrhopetalum 22
Cleisomeria 40-41
Cleisomeria lanatum 40, **41**
Cleisostoma 34-39
Cleisostoma discolor 34, **35**
Cleisostoma scortechinii **36**
Cleisostoma subulatum 36, **37, 38**, 39
coconut husk 114
coelogyne 3
collector 5
column-foot 6
column 6
cork slab 114
corybas 2, 5
Corymborkis 5, 42-43
Corymborkis brevistylis 42
Corymborkis rhytidocarpa 42
Corymborkis veratrifolia **42, 43**
cryptostylis 2
cultivation 112
Cymbidium 44-49, 115
Cymbidium atropurpureum 44, **45**

Cymbidium cloranthum **46**
Cymbidium dayanum 47
Cymbidium finlaysoniamum 44, **45**, 47 ,49
Cymbidium lancifolium **48**
Cymbidium simulans **49**

Day 47
Dendrobium 3, 23, 50
Dendrobium crumenatum 3, 4, 113
desmotrichum 50
Dimorphorchis lowii 5
diversity 5
dormant 2
dry spell 113

ecological niche 3
elephant orchid 87, 88
endemic 85
epiphyte 1
epiphytic orchid 3
establishment 113
eria 23

fern root 114
fertilization 6
fertilizers 116
Finlayson 47
Flickingeria 50-51
Flickingeria fimbriata **51**
foxtail orchid 87
fragrance 5, 107
Fraser's Hill 11

galeola 2
Garay 34, 110
Gaudichand 89
genera 1
Geodorum 55-57
Geodorum citrinum 55, 57
Geodorum densiflorum **55, 56, 57**
Geodorum purpureum 57
giant orchid 5, 54
goodyera 1
Grammatophyllum 52-54
Grammatophyllum papuanum 54
Grammatophyllum peliiflorum 52
Grammatophyllum speciosum 5, **52, 53**, 54
Griffith 62
grouping 1
Gunung Ayam 67
Gunung Blumut 67
Gunung Jerai 2, 3, 67
Gunung Tebu 67

Habenaria 1, 2, 58-61
Habenaria carnea 58
Habenaria medioflexa **58, 59**
Habenaria rhodocheila 58, 60, **61**

Habenaria trichochila 59
habitat 3, 112
Hammer 64
Hawkes 50
Holttum 10, 11, 13, 22, 24, 32, 33, 50, 52, 58, 62, 64, 85, 90, 97, 99, 104, 110
Hunt & Summerhayes 50

index Kewensis 28, 40
indigenous 1

jewel orchid 2, 11

keel 54
King, G. 62
Kingidium 62-63
Kingidium deliciosum 62, **63**, 115
kingiella 62

labellum 6
lady's slipper 66
Lim Swee Lim 32
Lindley 22, 40, 50
Linnaeus 52
lip 6
liparis 1
Loureiro 83
ludisia 5

maintenance 116
malleola 64-65
Malleola insectifera **64,** 65
mariposa 70
mentum 16, 51, 103
mid-lobe 6
monopodial 3
moth 70
mottling 66
Mount Ophir 3, 67

ovary 6

Paphiopedilum 1, 2, 6, 66-69
Paphiopedilum barbatum 66, 67, 68, 69
Paphiopedillum niveum **66**
Papilionanthe hookerana 3
pecteilis 1, 58
Pecteilis susannae **2**
Penang Hill 2, 67
pendulum 6
perfume 107
peristylis 1, 58
pesticide 116
petal 6
Pfitzer 106
phaius 1, 2, 116
Phalaenopsis 70-74
Phalaenopsis amabilis 70

Phalaenopsis cornu-cervi 6, **70, 71**
Phalaenopsis decumbens 62
Phalaenopsis fuscata **72**
Phalaenopsis gigantea 5
Phalaenopsis violacea 5, 72, **73, 74**
platanthera 58
pod 6
podochilus 4, 75-77
Podochilus tenuis 75
Pomatocalpa 77-80
Pomatocalpa kunstleri **76**, 77, 78
Pomatocalpa latifolium **78**
Pomatocalpa setulense 78, **79**
Pomato calpa spicatum **80**
Porphyrodesme 81-82
Porphyrodesme elongata 5, 81, **82, 83**
Porphyrodesme papuana 81
pouch 66
Pulau Langkawi 67, 91, 92

queen orchid 54

rainfall pattern 113
Reichenbach 22, 47
Renanthera 83-84
Renanthera coccinea 83
Renanthera elongata 83
Rennanthera histrionica **82**, 83, 85
Renanthera matutina 83
Renatherella 85-86
Renantherella histrionica 83, **84**, 85, 86
resting 113, 116
Rhynchostylis 87-88
Rhynchostylis gigantea 87, 88
Rhynchostylis retusa 5, **86**, 87
Ridley 28, 34, 85, 88, 99
Robiquet, P. 89
Robiquetia spathulata **89**
rock 3
Rolfe 59

saccolabium 34
sandal of venus 66
saprophyte 1
sarcanthus 34
Schlechter 22
seasonal variations 113
seed pod 6
Seidenfaden 28, 50, 57, 59
sepal 6
side-lobe 6
Smith 22, 62

Smitinand 90
Smitinandia 90-91
Smitinandia micrantha **90, 91**
Spathoglottis 2, 60, 92-96
Spathoglottis affinis **94**, 95
Spathoglottis aurea **95**, 96
Spathoglottis hardingiana 92
Spathoglottis plicata **92, 93**
species 1
spur 6, 18, 60
staminode 66
stigma 6
sugar-cane orchid 54
Swartz 44
Sweet 62
sympodial 3
synesepalum 6, 66

taeniophyllum 5
tainia 2
Taman Negara 3
terrestrial 1, 58
Thecostele 97-98
Thecostele alata **97**, 98
Thecostele secunda 97
Thours 23
Thrixspermum 99-100
Thrixspermum calceolus **99, 100**
tiger orchid 54
Trichoglottis 3, 5, 101-105
Trichoglottis brachiata 101, 105
Tricholglottis cirrhifera 101, **102**
Trichoglottis fasciata 102, **103**
Trichoglottis lanceolaria **103**
Trichoglottis misera **104**
Trichoglottis retusa 105
tuber 1
Turill 59

uncifera 110

Vanda dearii 6
Vandopsis 106
Vandopsis gigantea 106
Vanilla 2, 5, 107-109
Vanilla griffithii 107, 108, 109
Vanilla pilifera **108**, 109
Vanilla planifolia 107
vanillin 107
variation 5
velamen 1
Ventricularia 3, 110-111
Ventricularia tenuicaulis **4**, 110, **111**